# JANE'S AIRCRAFT SPECTACULAR
# PHANTOM

Text by BILL SWEETMAN          Illustrations by JAMES GOULDING

JANE'S

Copyright © Jane's Publishing Company Limited 1984

First published in the United Kingdom in 1984 by
Jane's Publishing Company Limited,
238 City Road, London EC1V 2PU

ISBN 0 7106 0279 0

Distributed in the Philippines and the USA and its
dependencies by
Jane's Publishing Inc.,
135 West 50th Street,
New York, NY 10020

All rights reserved. No part of this publication may be
reproduced, stored in a retrieval system, transmitted in
any form or by any means electrical, mechanical or
photocopied, recorded or otherwise without prior
permission of the publisher.

Typeset by D. P. Media Ltd,
Hitchin, Hertfordshire

Printed by Toppan Printing Co Ltd,
Tokyo, Japan

James Goulding would like to thank the following for their
help with research material for illustrations in this book:
Karen Stubberfield (McDonnell Douglas), Barry Wheeler
(Ministry of Defence) and Martin Horseman.

An F-4J of Marine squadron VFMA-333 drops a full load of retarded bombs. Note the auxiliary cooling passage open above the laden fighter's engine bays. (MDC)

## Introduction

Malevolence expressed in aluminium: words like these immediately spring to mind when you contemplate the brutish and formidable McDonnell Douglas F-4 Phantom. There are other possible descriptions: the most successful military aircraft of the post-1945 era; the only aircraft which in itself accounted for a complete generation in fighter development. But the looks of the Phantom tell their own story.

Start with the nose, drooped towards the ground as if to sniff out the Phantom's prey. Below the radome is the gaping smoke suppressor for the M-61 cannon. Behind it the fuselage sweeps up into vertical, precipice-like sides topped by the cockpit canopies. Behind these the form changes abruptly once again, with the flared, cavernous intake tunnels complete with power-shovel boundary-layer ramps. The fat centre-section perches on a broad slab of thin-section wing, which itself embodies some dramatic angles. The whole leading edge extends forwards and droops away from the main structure, parts of the trailing edge hinge down more than halfway to the vertical, and the tips are angled sharply upwards.

Behind the wing, matters get worse. Where normally one would find the rear fuselage, there is a skeletal spine in bare metal projecting beyond the raw stumps of the jetpipes. From this assembly the stabilizers droop as if to meet the wingtips, and the shape is further complicated by a mighty butcher's hook hinged to the underside.

The fact that this jagged machine was by a large margin the best thought-out fighter design of its day was not appreciated by a world which looked back to the smooth lines of subsonic streamlining and looked forward to dart-like, elongated Mach 3 profiles. It was a couple of years after its unveiling that the Phantom showed an astonished world what it could do, becoming the first aircraft to hold simultaneously the absolute records for maximum speed at high and low level, absolute records for time to height, and both absolute and sustained altitude records. The world put its eyeballs back in place, but the Phantom went on to show its other side. Throughout the 1960s and early 1970s the Phantom was the only Western fighter in service with a radar/missile combination which functioned at all effectively in a tactical environment. In both air-to-air and air-to-ground warfare it carried a far heavier armament than any of its contemporaries. It carried the first of a new breed of fighter radars and, on its multiple hardpoints, sported more new weapons and subtle electronic systems that ushered in a new era in air warfare.

The Phantom was not perfect, however. It had its drawbacks, particularly when it was used in roles for which it was not designed. The hulking fighter's worst problems were its mean behaviour in the stall and spin, and hefty subsonic drag. One fault was apparent in air-to-air close combat, the other in heavy-laden low-level long-range strike, and the Phantom was intended for neither of those missions. The important fact was that right up to the mid-1970s it was nonetheless better than anything else available.

Possibly the most telling testimony to the importance of the Phantom's design is the aircraft which was designed as its primary US Air Force replacement: the F-15

With its broad swept wing, cutback rear fuselage and aggressive aspect, the F-15 Eagle is instantly recognizable as the F-4's offspring. (MDC)

Eagle. The Eagle is very similar in size and shape to the Phantom, and carries a comparable armament. Generally, the two designs differ sharply where the designers have taken the weak areas of the Phantom and turned them into the major attributes of the Eagle. Although the Eagle is an entirely contemporary design, the positive and negative philosophical influence of the Phantom is tremendously important.

And yet the most remarkable thing about the Phantom story is that so little of it was planned. The development and career of the Phantom originated in a series of repeatedly changed and confusing official requirements. The design evolved in a jungle of detail changes, new technical additions and fearsome stability and control problems, while the Navy Bureau of Aeronautics seemed to delight in setting stiffer and stiffer requirements at every stage. Once development was complete, a second set of random factors took over. All the Phantom's contemporary rivals were single-mission aircraft; those that were not were scrapped during development. Then all the aircraft that should have supplanted the Phantom in the 1960s hit trouble in the form of cancellation, cost escalation and delays. This is truly a story of an aircraft that was successful in spite of everything that defence planning could do.

## It started in St Louis

Between 1930 and 1945 the US Army and Navy almost completely changed the sources from which they bought aircraft and set a pattern which, with few changes, has continued ever since. Virtually every aircraft bought by the Army, Air Force and Navy since that time has been built by a contractor which delivered aircraft in large numbers during the 1939–45 war. The exception to the rule is also the world's most successful military aircraft supplier – the McDonnell company or, as it has been known since 1967, the McDonnell Aircraft Company division of the McDonnell Douglas Corporation. It was the Phantom II that made the company an industry giant, but

McDonnell had already done well to get to the stage of building the first prototype.

James S. McDonnell – "Mr Mac" – formed his new company in St Louis in mid-1939. His choice of a site was far-sighted to the point of prescience. Little over two years later America was preparing to fight a war with the aid of an air force vaster than any pre-war imagining. To this day, building aeroplanes from riveted aluminium is a job which takes a great many man-hours, and not even the combined workforces of the West Coast's aircraft industry and the north-east's car companies could meet the demand for airframes. So the aircraft industry sent the factories to where the workers were: the mid-West, Kansas and Missouri, where farm mechanization had produced a surplus of labour. McDonnell was there already, and was soon busy with sub-contract work.

At the same time, though, Mr Mac was working on a more ambitious project: a long-range, heavily armed fighter. The Army Air Corps ordered two prototypes in 1941, but the project was an inauspicious start to the company's career, never enjoying very high priority. The first XP-67 did not fly until January 1944, the Continental engines and their turbochargers gave trouble, and the type was too late to be of significance anyway.

Before the XP-67 had made its first flights, McDonnell had been given a much more important contract: to design and build the USA's third jet fighter, the first propellerless aircraft to be designed for naval operations. This flew

First fighter from McDonnell was the XP-67, oddly reminiscent of the Rockwell B-1 with its wing/body blending and mid-set tail. Aerodynamically promising, it was let down by an inadequately developed powerplant and remained one of many bypassed prototypes of the era. (MDC)

as the XFD-1 – for some reason the Navy had assigned McDonnell the same "D" manufacturer designation as Douglas – in January 1946. Because of the relative quiet of the jet engines, McDonnell named the new fighter "Phantom".

Only 60 Phantoms were delivered, but the type became the first pure-jet aircraft to operate from a US carrier. It was in other ways thoroughly conventional, with a big wing for low-speed carrier landings and take-offs. Its main significance for McDonnell was that the basic configuration was sound and could easily form the basis for a scaled-up aircraft with the new and more powerful Westinghouse J34 engine. First flown as the XF2D-1, the resulting type entered service as the F2H-2 Banshee. When the Korean War broke out in 1950 the straight-winged Banshee and Grumman F9F Panther were the best fighters available to the US Navy, and were accordingly ordered in sizeable quantities. One of the Banshee's significant achievements was that it was the first operational single-seat jet fighter to be equipped with a radar fire-control system, Westinghouse's APQ-41. This was the first association of two of the partners which would form the Phantom II team.

Though Navy fighters were the bulk of McDonnell's business up to the late 1950s, the company continued to explore other lines of business ranging from missiles to helicopters. In August 1945 McDonnell responded to an Army Air Force requirement for a long-range jet fighter – because of the high fuel consumption of the jets, none of the new fighters had anything like the range of the piston-engined P-51, P-47 or P-38 – and won a contract to build two prototypes under the designation XF-88. The first flew in October 1948 and was named Voodoo in McDonnell's newly established supernatural tradition.

The Voodoo's design had been driven by the need to provide a great deal of internal fuel capacity to meet the range requirement. The swept wing was fitted with leading-edge flaps, and McDonnell's own afterburners were fitted so that the J34 engines – chosen because they were the right size for cruise performance – could get the heavy aeroplane off the runway. Fuel was carried in the stout fuselage and in wing tanks formed by the integrally machined skins. A McDonnell trademark made its first appearance: ducking the problems of airframe cooling around the hot afterburners, the designers hacked away the rear of the fuselage apart from a stump above the jetpipes which carried the tail surfaces.

The trouble with the XF-88 was that it was over-specified. On dry (unreheated) thrust the aircraft was under-powered, while with reheat working the fuel consumption was too high. What's more, the idea of providing escort fighters for the 10,000-mile-range B-36 had already been thrown out as impracticable by the time the XF-88 flew.

Some of the new type's features went into McDonnell's next fighter, the Navy's F3H Demon, which resembled the XF-88 in having swept wings, a portly fuselage and large internal fuel capacity. Like the later Banshee it was a single-seat all-weather fighter, and proved a reliable if not epoch-making design once horrible engine problems had been solved by drastic means. (The manufacturer was fired, and a new powerplant installed in all operational production aircraft.) The type was significant in another way: from the late 1940s the Navy and Air Force had been experimenting with

Carrier flying, and carrier-based aircraft, present special challenges. Canopy cracked open in case of problems, one of the Jolly Rogers' Sparrow-equipped F3H-2Ms quits the deck. (MDC)

The hulking F-101A Voodoo pointed the way to the F4H-1. (MDC)

guided air-to-air missiles which would in theory eliminate the need for guns and make kills almost certain. The Navy's engineers produced two missiles: Sidewinder, a small, simple weapon which homed on the heat from the target's jetpipe, and a larger, longer-range radar-guided weapon. Raytheon developed the first effective version of the radar-guided weapon, the Sparrow III, and it was first carried in service by the F3H-2M Demon in August 1958.

To some extent the Demon was the starting point for the design which became the Phantom II. A more significant influence, however, was the aircraft which resulted from the revival of the XF-88 and which shared its name. When it flew in October 1954 the massive F-101A Voodoo was twice as powerful as the next biggest fighters around. It also had by far the greatest internal fuel capacity of any fighter. Completely different in design from the XF-88, the new Voodoo shared its basic configuration. But with far greater thrust and the first variable-area intakes on a production aircraft, it was highly supersonic – briefly holding the world air speed record in 1958 – and could carry a heavy payload. Unfortunately, the basic Strategic Air Command requirement for the aircraft was scrapped just before the first flight. Tactical Air Command bought some Voodoos for the strike role but was committed to the Republic F-105, with its purpose-built nav/attack system, in the longer term. More Voodoos were used for reconnaissance, and even more were built as two-seat long-range interceptors for Air Defense Command. In fact the little remembered Voodoo was the first multi-role supersonic fighter and was eventually built in respectable numbers. In its day, however, it was regarded as something of a monster. When the Voodoo was announced in late 1954 most people were astounded by its size. They would have been totally baffled, confused and probably plain incredulous if they had been told that McDonnell was well advanced with the design of a fighter that would not only be heavier than the Voodoo but would operate from a carrier. But that was exactly what was proceeding across the St Louis drawing boards.

## Heavyweight champion

In 1952 the Navy gave Chance Vought and Grumman contracts for its first supersonic fighters. Possibly for this reason, McDonnell started looking at ways to meet the service's requirement for attack aircraft to replace the piston-engined Skyraider. The starting point was the Demon, a heavy aircraft with plenty of fuel capacity, and the first designs were designated F3H-G. However, the new aircraft soon lost all resemblance to its predecessor, taking shape as a large Mach 1.5 strike aircraft with a secondary interception role. It was fitted with two Wright J65s – developed from Britain's Sapphire – and carried four cannon, radar and a load of external weapons. In late 1953 the Navy Bureau of Aeronautics (BuAer) issued a development contract under the designation AH-1, signifying the primacy of the attack role, and in October 1954 two prototypes were ordered.

By May 1955 McDonnell was on the point of starting construction when BuAer told it to recast the AH-1 as a high-supersonic interceptor carrying a two-man crew and Sparrow missiles, and designed to provide air defence at a 290-mile radius, outside the fleet's own radar coverage. There were a number of reasons for the change, including the rapidly growing strength of the Soviet jet bomber force and the inability of the smaller Crusader and Tiger to carry a powerful radar and Sparrows. At the same time, BuAer ordered a scaled-up, high-supersonic, Sparrow-armed Crusader variant from Vought.

The McDonnell project was revised as the XF4H-1, a basically conventional design apart from its weight and remarkable structural density. While earlier fighters had consisted of an airframe with components arranged inside, the XF4H-1 was to be built around a strong, fuel-carrying keel to which were attached the nose, wing, engines and tail, with the skin shrink-wrapped around the carefully packaged components. This was essential, because the Navy wanted the whole aircraft to be no larger than the Demon – to fit the lifts on its medium-sized carrier – even though the earlier type had only half the power. The wing was in fact just fractionally bigger than that of the Demon, and to keep landing speeds reasonable the leading and trailing-edge flaps were to be "blown" by air bled from the engines. The thin sheet of high-pressure air energized the flow over the flaps and allowed the trailing-edge surfaces to be deflected to 60° down without flow breakaway. At this stage both the wing and the tailplane were flat.

As design of the XF4H-1 got under way, another major change hit the programme. Since 1945 the Navy had bought most of its jet engines from Westinghouse and Pratt & Whitney, with some from Allison and Wright. General Electric, builder of the first jet engines to fly in the USA, had not sold any engines to the Navy and had been enjoying mixed fortunes with the Air Force. After seeing its wartime designs handed over to Allison, GE had bounced back with the J47, produced in massive numbers for Sabres and B-47s, but had then been wiped out by Pratt & Whitney's J57. A project team under Gerhard Neumann had been set up in 1951 to produce a counter-weapon. The result was the "variable-stator" engine, with movable vanes between the compressor stages so that the engine could adapt itself to the different conditions of starting, subsonic and supersonic flight. In particular, the engine could run at high pressure ratios in supersonic flight, developing more thrust for the same airflow and, all other things being equal, for a slimmer and less draggy engine and intake. In late 1952 a GE variable-stator engine was chosen as the powerplant for the US Air Force's new supersonic bomber, the Convair XB-58, and was designated J79.

The J79 ran in June 1954 and soon started to attract serious interest from fighter designers. Lockheed redesigned the F-104 Starfighter around the engine, and in 1956 Grumman fitted J79s to a pair of modified Tigers, designated F11F-1F. The Super Tigers were truly transmogrified, attaining not far short of Mach 2 and climbing above 75,000ft, and for the first time the US Navy took notice of GE. The slim shape of the J79 – an essential for the podded installations of the B-58 – was a tremendous boon to the F4H with its side-by-side engines, reducing cross-section and wave drag. So was the excellent performance of the engine at supersonic speed. In all, the J79's lower drag and higher thrust promised to transform the F4H from a Mach 1.5-plus aircraft into a Mach 2-and-then-some sprinter. It was adopted for the new fighter before construction started in August 1956.

Other features of the design were associated with the high design speed. The Sparrow missiles were semi-recessed, a far easier solution than internal stowage yet offering much less drag than pylon mountings. (This McDonnell idea was finally adopted by other designers in the late 1970s.) In a decision which was later regretted, BuAer dropped the requirement for built-in cannon armament. The air intakes also called for a great deal of design work. The original AH-1 had simple fixed cheek-type inlets which would have become rapidly less efficient at speeds above Mach 1.5. The boundary layer along the fuselage side would also become thicker and more turbid with increasing speed. So the inlets were moved outboard and fitted with big entry ramps. The lip of the ramp deflected the primary shock wave ahead of the inlet mouth, and the ramp could move to position the shock in line with speed changes. In the inlet itself, just ahead of the engine, there was an annular valve which spilled excess air down the outside of the J79, cooling accessories, providing extra thrust in the afterburner and creating a thermal barrier between the hot engine and the fuel tanks in the spine. The powerplant section terminated in a variable exhaust nozzle: at full power the exhaust nozzles opened and the engine airflow filled the space around the skeletal tailboom, reducing base drag. In summary, the Phantom's powerplant was ideally integrated

Hammered out during a long and difficult development, the aerodynamic features of the XF4H-1 proved highly successful and needed little revision. The first aircraft flew with all the "fixes" in place. Note the extended top lip of the prototype's inlet. (MDC)

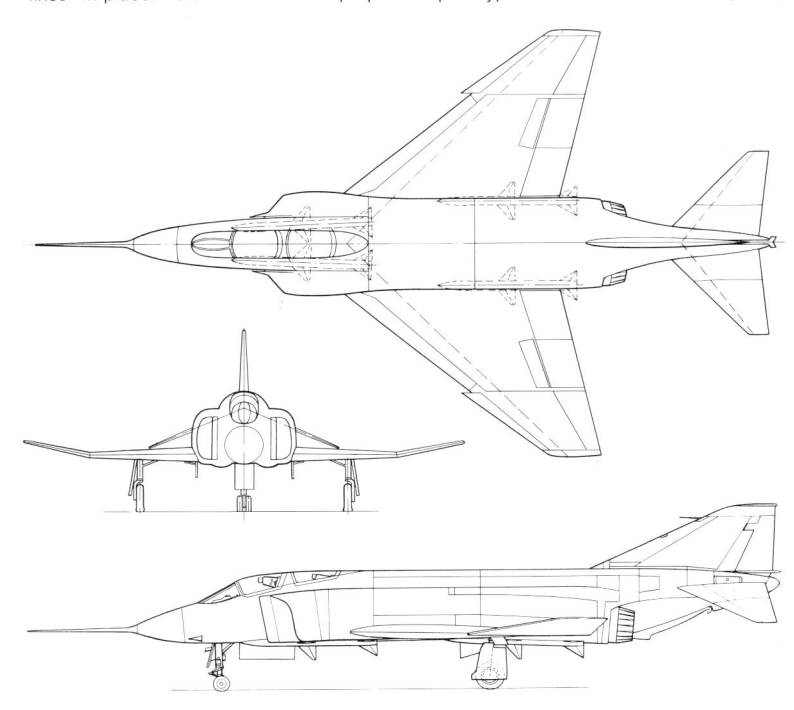

This clean, straightforward aircraft is the mock-up of the AH-1 single-seat attack type. Note the straight wing and tail and the plain inlets. (MDC)

The final intake design used this large hinged, perforated ramp. (Jay Miller)

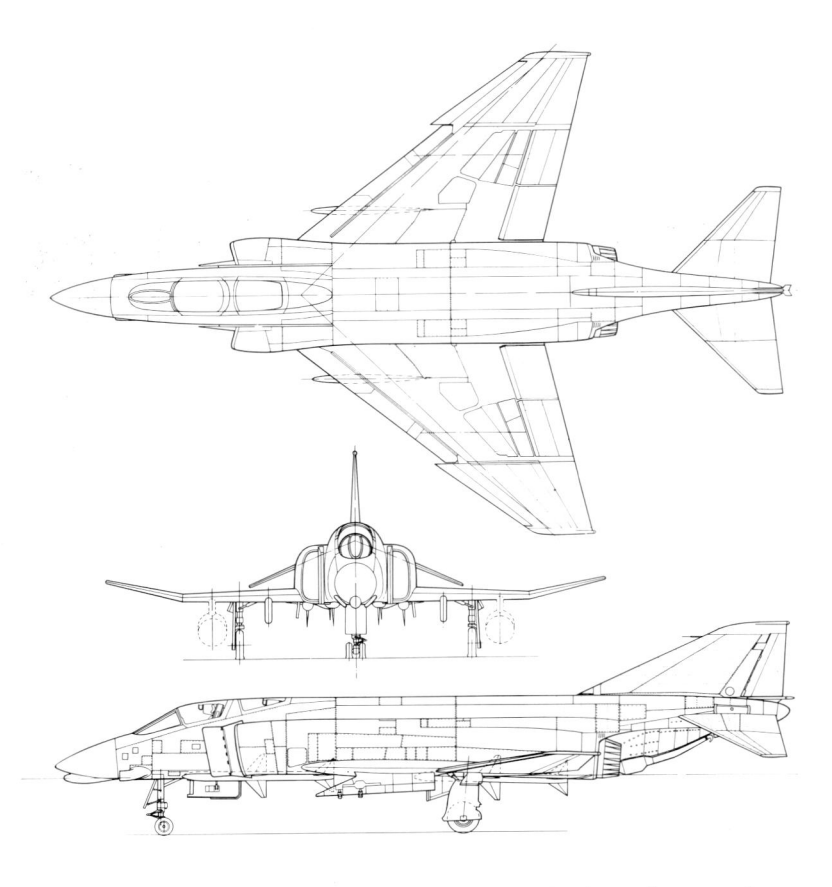

Early F4H-1 development aircraft featured a lower-set rear seat and a smaller nose radome. The final inlet design was introduced early in the test programme. Visible here is one of the most important features of the design – the "semi-conformal" missile installation. (MDC)

with the airframe to provide high supersonic thrust for minimum cross-section and weight.

Most of the new aircraft was made of aluminium, although the central spine was largely steel and titanium. The design of the tailcone presented interesting problems because of its proximity to the 1,600°C exhaust from the afterburners. The lower surfaces were given a double air-cooled skin covered with an outer layer of ceramic-coated titanium "shingles". For the same reason the inner section of the stabilizer had to be made from heat-resistant metals and alloys similar to those used in jet engines.

But powerplant and structural design were simple compared to the problems of stability and control. The Navy had demanded that its new fighter should have better subsonic and supersonic stability and handling than its first generation of supersonic fighters, but wind-tunnel tests of the McDonnell XF4H were showing results which varied from the mediocre to the horrible. Violent pitch-up was present at low and high speeds, and in high-speed turns the aircraft was liable to roll/yaw coupling: it was unstable in turns, with a tendency for rapid rolls to make the aircraft diverge in yaw under gyroscopic forces. Worse still, the Phantom's skittishness affected all three axes, so solutions had to be found to each particular form of instability without making another one more severe. In 5,000 hours of wind-tunnel testing – an enormous total for the 1950s – the company tried 100 tail configurations and 75 wing planforms before the gremlins were pronounced exorcised. The job was done so well that no aerodynamic modifications were needed as a result of flight tests, and only a few were called for in later development.

The low-speed pitch-up was a result of tip stall: the tips of the swept wings were stalling before the roots, and as the tips were aft of the centre of gravity this caused the nose to pitch up. The cure was to add area at the tips by extending the chord forwards on the folding outer panels. The resulting dogtooth also generated a vortex which slowed down the spanwise-and-aft flow responsible for the tip-stalling.

Pitch-up at high speed was caused by downwash over the tailplane in the transonic regime. It could have been cured by using a cruciform or T-tail, but this would have left McDonnell with the need to add some vertical area to restore directional stability. So the designers drooped the slab tail by 23° so that it would penetrate the downwash zones and operate in normal airflow, and at the same time add some side area to help the yaw-roll problem.

The final fix deemed desirable was a 3° dihedral (upwards angle from the centreline) on the wing to improve roll stability and finally eliminate the yaw-roll coupling. But by this time tooling for production was well advanced, so the designers achieved the same effect by applying 12° dihedral to the outer, folding wing panel, leaving the main wing structure unchanged. All the aerodynamic modifications – the cranked wings, the dogtooth and the drooped tailplane – were incorporated on the first prototype before it flew.

As noted above, the F4H was one of the first aircraft to be designed with blown flaps, which in this case extended across the entire leading edge of the wing and the inboard trailing edge. The ailerons filled the mid-portion of the

**Above** The first Phantom over the Missouri river on an early flight. Marvin Marks, one of the design team, later recalled looking at the jagged shape and thinking: "My God, what have we created?" The "X" prefix was not applied to the aircraft. (MDC)

**Below** Weightlifting contests between the Navy and the Air Force were a feature of the early 1960s. Here, in April 1961, an F4H-1F development aircraft demonstrates its ability to lift 22 Mk 82 500lb bombs. (MDC)

Section of Marine Corps F-4Js rockets towards the top of a formation loop. (MDC)

James Goulding

Reconnaissance crews are always the unsung heroes, and the exploits of RF-4C pilots in Vietnam tended to follow the pattern. The RF-4C is currently the main front-line tactical reconnaissance aircraft in US service. This aircraft belongs to the 30th Tactical Reconnaissance Squadron, part of the 10th Tactical Reconnaissance Wing at RAF Alconbury, England. The RF-4C was the first reconnaissance version of the Phantom, flying in August 1963. It outlasted the F-4C and F-4D in production, and 505 were built before production ended in December 1973.

**RF-4C leading data**

**Mission** Tactical reconnaissance

**Powerplant** Two General Electric J79-GE-15 turbojets rated at 10,900lb (4,944kg) dry thrust and 17,000lb (7,711kg) with afterburning. Cartridge starting

**Armament** None

**Avionics** APQ-99 forward-looking mapping and terrain-avoidance radar; ASN-56 inertial navigation system (being replaced by Lear Siegler ARN-101); AAD-5 infra-red detector; APQ-102 side-looking airborne radar. Some aircraft now being fitted with improved Slar and Terec (tactical electronic reconnaissance) sensor for detecting emitters

**Dimensions**
**Wing span** 38ft 5in (11.7m)
**Length overall** 61ft 0in (18.59m)
**Height overall** 16ft 3in (4.96m)
**Wing area** 530ft$^2$ (49.2m$^2$)

**Weights**
**Empty** 28,000lb (12,700kg)
**Max take-off** 54,600lb (24,765kg)

**Performance**
**Maximum speed** Mach 2.2
**Service ceiling** 55,000ft (17,000m)
**Combat radius** 780nm (1,450km)

James Goulding

US Navy pilots outdid the Japanese pilots of 1941-45 when it came to decorating their aircraft. This F-4B was in action off Vietnam with USS Coral Sea (CVA-43) in mid-1972. The multi-coloured "Supersonic Eagle" emblem was carried by the fighters of Air Wing 15. Later it was realized that the gaudy paint jobs were an unnecessary aid to the enemy. The F-4B was the standard production version for the US Navy and Marines until it was replaced in 1966 by the F-4J. A total of 649 were built. Some have been converted into QF-4B drones and others updated as F-4Ns.

## F-4B leading data

**Mission** Carrier-based air superiority and combat air patrol

**Powerplant** Two General Electric J79-GE-8 turbojets rated at 10,900lb (4,944kg) dry thrust and 17,000lb (7,711kg) with afterburning. Air starting

**Armament** Normally four Raytheon AIM-7E Sparrow III recessed into fuselage, four Raytheon/NWC AIM-9D Sidewinder on two dual wing pylons

**Avionics** Westinghouse APQ-72 radar fire-control system

### Dimensions
**Wing span** 38ft 5in (11.7m)
**Length overall** 58ft 3in (17.76m)
**Height overall** 16ft 3in (4.96m)
**Wing area** 530ft² (49.2m²)

### Weights
**Empty** 28,000lb (12,700kg)
**Max take-off** 54,600lb (24,765kg)

### Performance
**Maximum speed** Mach 2.2
**Service ceiling** 55,000ft (17,000m)
**Combat radius** 780nm (1,450km)

trailing-edge span, the rear of the folding wing section being fixed, and were arranged to droop 10° with the flaps extended. Nearly all the rolling force was in fact provided by lift spoilers above the wing, while at low speeds the ailerons and rudder were automatically interconnected to steer the aircraft through a co-ordinated turn. The complex high-lift and control system was intended to allow the heavy aircraft to stay airborne at 118kt, well under a tenth of its design maximum speed.

While the F4H was not quite as remarkable in terms of sheer performance as the near-contemporary Vigilante, it was nevertheless an ambitious design. Its wing loading was high for a carrier-based fighter, it was intended to carry a very heavy armament, and the complexity of its flight controls and powerplant far exceeded that of any previous naval fighter. The crew of two were to be assisted by a range of sophisticated avionics with which they shared a tightly packed forward fuselage. The heart of the system was an AiResearch central air-data computer, an analogue processor which provided all the other systems with the essential information which they needed. One important subsystem was the General Electric automatic flight-control system, which not only performed the normal functions of an autopilot but also acted as a three-axis autostabiliser. (The aircraft was however designed so that it could be flown safely with the autostabilisation inoperative.) Another vital system controlled the movement of the powerplant ramps, valves and variable-area nozzles according to aircraft speed, altitude and temperature.

The biggest single element in the avionics was the Westinghouse APQ-72 radar, the primary sensor in a fire-control system designed to find targets beyond visual range and then illuminate them with a steady "tracking" beam. The Sparrow missiles would then home on to the reflected radiation and destroy the target when it was still tens of miles from the interceptor. The APQ-72 was an improved version of the APQ-50, which had been proposed for the original AH-1 attack aircraft; successive versions of the basic 50-mile-range X-band radar were to equip the majority of production aircraft. Two other weapon-aiming devices were specified for the F4H: an infra-red detector, the ACF Electronics AAA-4, and the Bendix/Eclipse-Pioneer AJB-3A bombing system, which allowed the aircraft to perform a "toss" bombing attack, the standard method of delivering nuclear weapons. One respect in which the F4H weapon-aiming systems differed from those of earlier Navy fighters was that there was no provision for gun sighting.

The Navy attached enough importance to its new fighter requirement to order substantial development batches of both the McDonnell and Chance Vought contenders. The first into the air was the St Louis product, which flew on May 27, 1958, as the XF4H-1 Phantom II. Various names had been proposed and rejected before the company and the Navy opted to re-use the name of their first jet fighter. The Crusader III flew in the following month but was clearly outclassed by its larger and more powerful rival. The Phantom II could carry more missiles; it had a crew of two, at that time considered desirable for a complex radar-missile interceptor; it had two engines, which in the eyes of a naval aviator

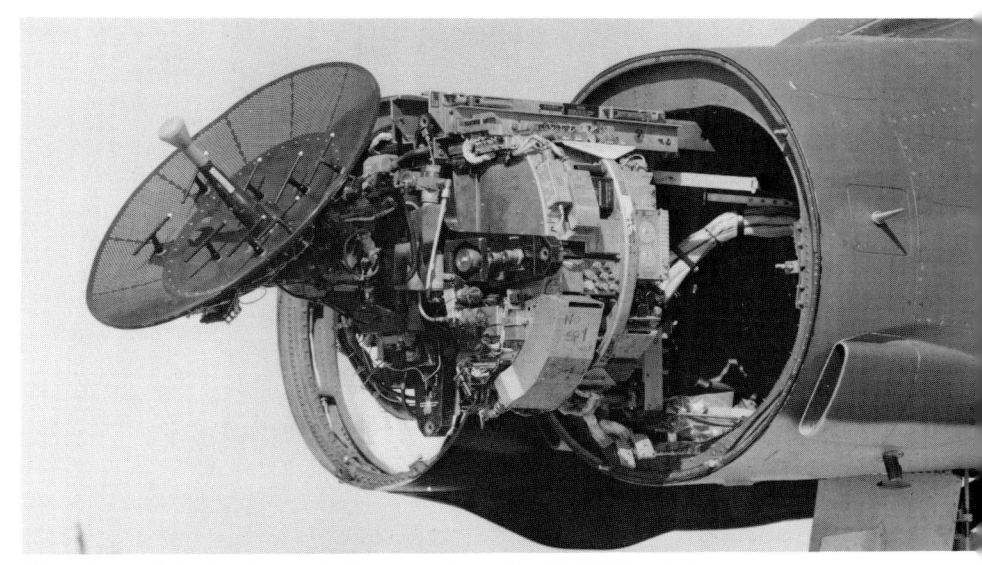

Westinghouse APQ-109 radar on an F-4D. Someone has written "NO SPY" on what was a very advanced set for its day.

is a cardinal attribute; and the J79 had wiped out the Crusader III's speed advantage. The Navy started evaluating the two aircraft as soon as flight trials were under way, announcing its verdict in December 1958: the Phantom was to be the Navy's next all-weather interceptor.

As McDonnell began preparations for switching its production effort from the Demon to the new aircraft – the last F3H was delivered in November 1960 – Phantoms began to roll out of the plant and join the test programme. Including the single XF4H-1, there were to be 23 aircraft assigned to the flight-test programme. They would be joined by early production aircraft, which also acted as replacements for machines lost in accidents.

Changes to the aircraft in the development phase were relatively few. The most visible was a redesigned front fuselage, dictated by altered requirements: the Navy wanted a better downward view for the pilot, so the nose was drooped slightly. It was also decided that it would be tactically useful for the back-seat crew member to have more than a token view from the cockpit, so the back seat was raised and a new bulged hood was installed. The final touch was added to the Phantom's unmistakable nose with the decision to modify the APQ-72 with a bigger reflector dish giving greater range and accuracy. The new radome bulged beneath the nose, while the IR scanner appended beneath it completed the now familiar droop-snoot look. Apart from some changes to the inlets, the nose changes were the only external alterations. Inside, the main distinguishing feature of early aircraft was the engine. The XF4H-1 had first flown with 14,800lb-thrust J79-GE-3s, as used on the F-104C. Later it was fitted with the more powerful (16,150lb-thrust) GE-2A, which was also used on all the other research and development aircraft and the first few "production" examples. The latter were not issued to operational units: once F4H-1s began to appear with the definitive 17,000lb-thrust J79-GE-8 the surviving GE-2 aircraft were redesignated F4H-1F and used for continuing trials.

These tests rapidly proved that the Navy's new all-weather fighter was something of a phenomenon. Accepted wisdom had always laid down that the modifications needed to operate an aircraft from the deck of a ship were inevitably detrimental to performance and efficiency,

17

and that the best the designer of a carrier-based aircraft could do was to reduce the margin of superiority enjoyed by its land-based equivalent. The Phantom II had clearly not heard about the accepted wisdom. The all-weather, radar-laden, two-seat Phantom was actually faster on the level and in the climb than the Air Force's F-104 Starfighter, which was then near-legendary for high performance. It was also as well equipped and as heavily armed as the Air Force's premier interceptor, the F-106, but was usefully faster and could land on a carrier rather than a two-mile concrete runway.

The US Navy was delighted, as was the manufacturer. Through 1960 and 1961 no small amount of effort was expended to demonstrate the performance of the Phantom with a long series of world records. At that time most of the absolute, time-to-height and closed-circuit speed records were held by the US Air Force or the Soviet Union. But there was more to the record attempts than simple inter-service rivalry or patriotism: they were closely connected to a political lobbying campaign to sell the F4H to the US Air Force.

Traditionally the two services had acquired completely different aircraft. The only recent exception had been the Navy's Douglas Skywarrior bomber, though the Air Force had subsequently revised the design to the point where its B-66A bore only a general resemblance to the original. But there were a number of reasons why the possibility of selling the Phantom to the Air Force could be seriously entertained by 1960.

One important reason was that the Phantom was unique in that it combined three important characteristics. It was fitted with a powerful radar and radar-guided missiles; it was highly supersonic, thanks to the great power of its engines (the production Phantom was more than 40 per cent more powerful than any Air Force fighter except the F-101); and because of its origins in the AH-1 long-range attack fighter it could lift a large external payload.

No other aircraft could offer this combination of capabilities because for the previous ten years the US Air Force had steered consistently away from the 1939–45 concept of a fighter as a high-performance aircraft equally capable of ground attack, reconnaissance and interception. Instead the service built specialized types which, although they carried the same F-series designation, were designed for completely different roles. By 1960 the US Air Force Air Defense Command was equipped with specialized high-altitude interceptors such as the Convair F-106 and F-102, which could not carry bombs or fire guns. Tactical Air Command was spending most of its money on the Republic F-105, a low-level, all-weather attack aircraft with only a gun for air-to-air combat. The same trend had been followed in other countries: an aircraft which was simply a "fighter" was regarded as an outmoded concept. To be accepted, a new aircraft had to be either a strike fighter biased towards ground attack, or an interceptor dedicated to the destruction of nuclear bombers.

Of the Phantom's contemporaries, two in particular could have offered some multi-mission capability. Both had a single large engine, were slightly smaller than the Phantom and, similarly, had adequate internal fuel capacity, provision for an effective radar-missile combination and potential for a large external payload. Both were scrapped: the Crusader III

Phantom and Sparrow were an inseparable team: a B from the first Phantom unit, VF-121, fires an AIM-7 on a practice mission. (MDC)

was beaten in a straight fight by the F4H and the British Hawker P.1121 was cancelled due to an overwhelming lack of interest from the Royal Air Force.

As the first Phantoms began to leave the production line in late 1960 another factor was beginning to be recognized. Between 1938 and 1948 the speed of service fighters had doubled. In the next ten years it doubled again, but by 1958–60 speed had reached a plateau defined by the limits of conventional jet engines and light-alloy structures. Higher speeds could be obtained, but only by specialized aircraft which were highly expensive and compromised in other performance regimes. At the same time more Air Force money was being diverted into missile programmes. By the end of 1960 there was little left of the Air Force's Mach 3 projects, and the requirement for a successor to the F-105 was written around a Mach 2.5 dash speed. The implication for McDonnell was that the F4H would not be ruled out of Air Force plans simply because it could not attain Mach 3, as might have been the case in 1958.

One last change cleared the way for the next stage of the Phantom's career: the election in November 1960 of President John F. Kennedy, leading an administration committed to change and new ideas. One of the most controversial members of the Kennedy cabinet was the Secretary of Defense, Robert S. McNamara. Former Ford executive McNamara and his team of defence theoreticians saw no essential difference between building Fairlanes and defending the United States. The way to success and victory over the competition was to get the best defence possible at the lowest price possible, no matter what traditions were thrown out along the way. Out went the Mach 3 bomber programme

Planned to replace the F-4 in USAF and Navy procurement in the late 1960s, the GD F-111 proved insufficiently versatile and too expensive.

despite efforts to resurrect it. An enormously expensive, decade-old white elephant, the airborne nuclear power project, was finally deprived of funding. But the most important decision for the Phantom programme concerned the long tradition by which the Navy and Air Force had used different aircraft. The generals and admirals tried unavailingly to convince McNamara that their needs were separate, and early in 1961 the biggest outstanding Air Force and Navy fighter requirements were combined under the designation TFX (tactical fighter – experimental). This decision indicated that the Department of Defense would favour any other moves towards "commonality" of equipment. In addition, the fortunes of the TFX programme were to affect profoundly the Phantom in a way that nobody, least of all McNamara's bow-tied "whiz-kid" computer drivers, could have expected in early 1961.

Between 1959 and 1962 McDonnell and the Navy mounted a series of successful assaults on an unprecedented range of world records. Project Skyburner took this F4H-1F to Mach 2.42 for a world air speed record. (MDC)

Project Top Flight was the name given to a September 1959 series of time-to-height attempts. Even with early engines installed, the big F4H-1 proved the equal of the smaller, lightly equipped F-104. (MDC)

## *Record collection*

1961 saw McDonnell and the US Navy engaged in two tasks. The first was the onerous and painstaking job of getting the new aircraft into service. The first squadron to receive the Phantom was VF-121, which started training in February 1961 at Naval Air Station Miramar, near San Diego in California. The second was VF-101 at NAS Oceana in Virginia. The aircraft was carrier-qualified in October 1961, following a four-day series of intensive trials aboard USS *Saratoga*. The Phantom was eventually cleared to operate from all the US Navy carriers except the smaller *Essex* and *Hancock* classes. By the end of the year Phantoms were replacing Skyrays with the 6th Fleet in the Atlantic and the 7th Fleet in the Pacific. The Navy then had only two fighters in production – the Phantom and the Crusader, the latter being needed to equip the smaller carriers – and a third, the TFX, at an early stage in development. Never before had the service been so reliant on a single aircraft type.

The second exercise had got under way in the preceding year and attracted far more publicity as the new fighter toppled world records like ninepins, starting with a zoom-climb altitude record in late 1959. It should of course be recognized that record performance figures are far removed from the sort of speeds, altitudes and payloads usually encountered in service. To begin with, a record attempt is always made by a specially selected pilot who has extensively studied and practised the ideal combination of altitude, flight profile and control setting for the particular mark being set. The aircraft is always in peak condition, having been carefully checked by the manufacturer's best people. Controlled conditions for the attempt are also important: maximum performance is obtained with minimum fuel load, so the record flight is made with a tiny reserve of fuel on a clear day with the lowest probability of bad weather. The aircraft will be cleaned up and may well be stripped of numerous items of internal equipment to save weight. Finally, the rules allow for almost any modification short of rocket boost, the limit being set only by the amount of cash available for the attempt. Not surprisingly, the performance achieved is a great deal higher than that which a fully equipped service aircraft can attain under normal operating rules, even without external weapons or fuel tanks. Record attempts are like official car mileage figures: they should be used for comparative purposes only.

Each of the record attempts had a message for the supporters of the Air Force's existing aircraft. The first marks to fall to the F4H-1 were the 100km and 500km closed-circuit speed records, set by an early development aircraft (with the original smaller nose) in September 1960. The 500km record, an average speed of 1,056kt, testified to the Phantom's combination of moderate supersonic drag and high internal fuel capacity. On the shorter circuit – equal to a sustained 3g, 8.6nm-radius turn – the Phantom attained 1,207.2kt, although the team claimed that the true speed was over 1,250kt (about Mach 2.25) because the aircraft went slightly wide of the circuit.

The more spectacular absolute records had to wait for the availability of the J79-GE-8 engine and the final tuning of the intake design. The first to fall, in the spring of 1962, were a series of time-to-height records. Earlier record-breakers had shown how altitude records could be set by climbing to medium altitude, accelerating on the level and pulling up into a ballistic climb until the combination of engine thrust, wing lift and momentum gave out. The "zoom climb" technique took the F4H-1 to 30,000m in 6min 11.43sec, with a string of time-to-height records falling along the way. The airspeed at the top of the climb was reported as 37.5kt, but the F4H had officially outclimbed the hottest Air Force fighter, the F-104.

Next, in August 1962, it was the turn of the USAF's premier strike fighter, the F-105, to be shown up. Pre-supersonic rules for speed records had demanded two runs over a 3km course at an average altitude of 100m or less. The last record set under those rules stood at 655kt and was resoundingly shattered by an F4H-1 with 753.64kt, or Mach 1.18. In the same month one of the F4H-1F evaluation aircraft set an unofficial but still important record: modified with multiple ejector racks on its fuselage and wing weapon stations, it lifted 22 standard 500lb bombs to register a total of 12,500lb of external ordnance (actual bomb weight was

The final records to fall to the F4H-1 were a series of time-to-height marks taken in early 1962 by a late development aircraft powered by the production-standard J79-GE-8 engine. (MDC)

An F4H-1F streaks across the California desert to set the first clearly supersonic low-level speed record, sustaining Mach 1.18 less than 100m above the ground. (MDC)

565lb), more than any previous fighter aircraft. The USAF retaliated, lifting 14,960lb (24 500-pounders) under an F-105D in October, but the strike potential of the F4H had been clearly demonstrated.

Meanwhile, the Navy and McDonnell prepared to tackle the high-speed, high-altitude regime. The USAF's standard interceptor, the F-106A Delta Dart, had taken the absolute speed record to 1,324kt or Mach 2.3 in 1959. The sustained altitude record went back further, having been set by a subsonic aircraft. (It was something of a dead letter, as Lockheed U-2s and Martin RB-57s broke it daily but were not cleared to make record attempts.) The one obstacle to further Phantom successes was the fact that the engines tended to overheat in sustained runs above Mach 2, as the inlets generated higher pressures and temperatures with higher speed. A system of pre-compressor cooling (PCC) was therefore designed. This sprayed an alcohol/water mixture into the inlet ducts ahead of the engines; the vaporization of the liquid extracted heat from the inlet air and reduced temperatures throughout the engine. In late November 1961 an F4H-1 fitted with PCC set a new world air speed record of 1,395kt or Mach 2.42 at 45,000ft over Edwards AFB. A few days later a similarly equipped aircraft sustained 66,443ft over a similar course at Mach 2.2, another world record. PCC was considered for production aircraft, but not actually adopted.

In service the initial Phantom versions could attain Mach 2 with four Sparrows in place and partial internal fuel, although external stores such as additional Sparrows or Sidewinders would lower the maximum. Supersonic flight at low level is not usual in service operation, normal speeds ranging from 450kt to 600kt depending on external load.

By late 1961 McNamara's experts were asking the US Air Force why it should not buy F4H-1s. The USAF was still asking for money to restart production of the F-106A – production had halted in late 1960 – and to complete a 1,400-aircraft, 14-wing tactical fighter force of F-105Ds. The DoD evaluated the Phantom against the F-106A and found that the Navy fighter had a better payload/range performance and 25 per cent more detection range in its radar, and that it needed 30 per cent fewer maintenance man-hours per hour in the air. The Air Force request for 200 more F-106s did not get very far. The USAF also evaluated the F4H-1 in ground-attack configuration. The Defense Department intensified its pressure to have the F-105 replaced with the Phantom, and in March 1962 it was announced that four wings of Phantoms would replace four wings of F-105s in the Fiscal Year 1963 purchase plan. Shortly afterwards the Air Force and DoD decided to reduce planned F-105 strength to seven wings and to terminate production of the Republic fighter in 1964. Three more wings of Phantoms would be ordered instead. Also ordered in March 1962 was a new reconnaissance version of the Phantom, Republic's RF-105 being cancelled. The standard fighter was to be designated F-110A and the reconnaissance aircraft RF-110A.

Labelled an F-110A, this aircraft is one of a few F4H-1s loaned to the Air Force in 1962 to speed the introduction of the Phantom. (MDC)

While the F-110As moved down the assembly line several Navy F4H-1s were loaned to the Air Force for familiarization and demonstration, promoting the rapid integration of the aircraft into US Air Force service. A more important timesaver, though, was the careful control by the Air Force and DoD of the number of changes to be introduced for Air Force service; there was to be no repetition of the Skywarrior/RB-66 story. Changes were very few: a version of the J79 fitted for cartridge starting, the GE-15, was installed, and the wing root was slightly bulged to accommodate bigger wheels and tyres. The rear-seat occupant was given a set of flight controls. The arrester hook and folding wings remained. The retractable refuelling probe of the Navy aircraft was replaced by a socket for an Air Force tanker's refuelling boom. The F-110A radar, the Westinghouse APQ-100, was a derivative of the APQ-72 with a different display, while the Air Force also installed a different bombing computer and a Litton ASN-48 inertial navigation system (INS). Some consideration was given to installing a twin-barrel 20mm cannon made by Hughes, but the penalty in weight and internal volume was excessive and the US Air Force instead started development of a podded version of the six-barrel M-61 Vulcan. The F-110A was also equipped to carry the Bullpup air-to-surface missile. It was probably fortunate that the Air Force expected to buy only four wings of F-110As before introducing the TFX. Otherwise, military procurement being what it is, they would most likely have taken the opportunity to alter the design beyond recognition.

Development of the RF-110A was a bigger task, involving the installation of a battery of cameras and electronic sensors in a longer and slimmer nose. The two prototypes were based on Navy F4H-1 airframes to speed development. The RF-110A carried no fire-control radar, just a simple APQ-99 weather and mapping unit in a smaller radome. Behind that were oblique and high and low-level panoramic cameras, and the sides of the fuselage contained miniature side-looking radar and infra-red detection equipment. The reconnaissance aircraft also had slightly less fuel capacity and a different INS. Shortly after development started the US Navy ordered a small batch of similar F4H-1P reconnaissance aircraft to equip Marine Corps squadrons.

Towards the end of 1962 commonality finally overtook the aircraft designation systems used by the USAF and USN, the cumbersome US Navy system being abolished completely. The designation F-4 was allotted to the Phantom. The remaining F4H-1Fs were designated F-4A, some dual control trainers being later referred to as TF-4As. The Navy/Marine F4H-1 and F4H-1P became the F-4B and RF-4B, and the USAF versions became the F-4C and RF-4C before the first examples flew.

The first YF-4C flew on May 27, 1963, exactly five years after the first flight of the XF4H-1. Deliveries to the Air Force started before the end of the year. The next version to

Air Force requirements launched the RF-4, later adopted by the Marines and export customers. (MDC)

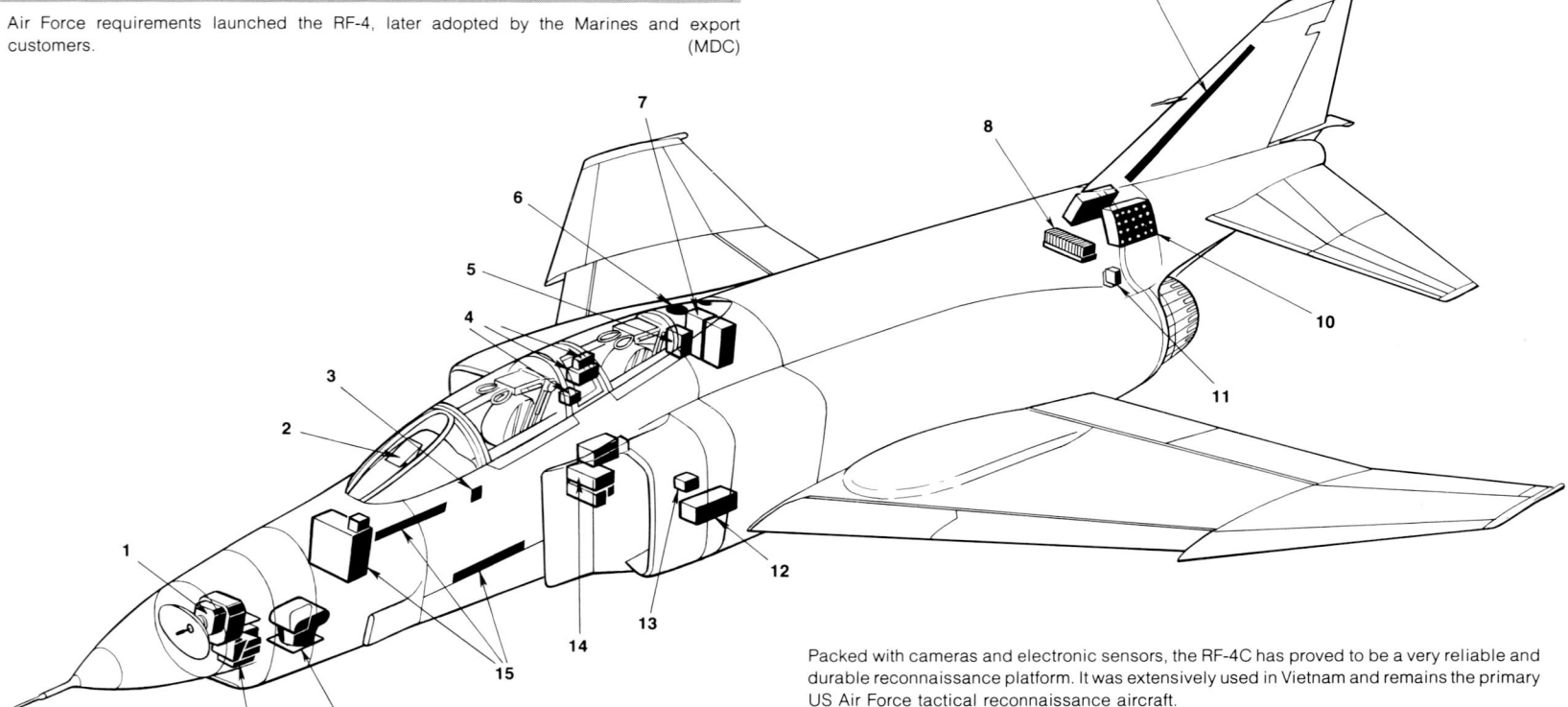

Packed with cameras and electronic sensors, the RF-4C has proved to be a very reliable and durable reconnaissance platform. It was extensively used in Vietnam and remains the primary US Air Force tactical reconnaissance aircraft.

1 APQ-99 forward-looking radar
2 LA-313A optical viewfinder
3 L-285A photo-flash detector
4 ASN-56 inertial navigation system
5 Communications/navigation/identification power supply, UHF navigation, RF navigation
6 IFF antenna
7 APQ-102R/T sideways-looking radar (SLR) antenna
8 ARC-105 HF transceiver
9 HF shunt antenna
10 Photo-flash ejectors
11 HF antenna coupler control
12 ASQ-90B auxiliary data-annotation system
13 Voice recorder
14 AAS-18A infra-red detection system
15 APQ-102R/T sideways-looking radar, recorder and antennae (alternative installations: one KS-87B camera with 18in focal length lens and LS-58A stabilised mount, or two KS-87Bs with 6in or 18in focal length lens and split-vertical mount)
16 KA-56D low-altitude panoramic camera with twin side oblique mount (or SLR recorder, or one oblique KS-87B with 12in or 18in focal length lens)
17 KS-87B forward-framing camera

appear was the long-nosed YRF-4C, which flew in August 1963, with the production-standard aircraft appearing in the following May. The first of the Marines' RF-4Bs was flown in March 1965.

One more version rounded out this first generation of Phantoms. This was an improved Air Force type, the F-4D, which replaced the F-4C in production in early 1966. Most of the changes were electronic, designed to provide better weapon-aiming. As on the RF-4C, some of the additional electronics displaced some of the fuel, occupying a bay where part of No 1 fuel cell had been. The new equipment included the APQ-109 radar – another version in the Westinghouse series, with added air-to-ground ranging ability, a better gunsight and lead-computing equipment – and an improved INS. The new radar had a slightly larger antenna in a bigger radome, and the under-nose IR sensor, which had proved of questionable worth, was deleted during production.

By mid-1966 well over 1,000 F-4s had been delivered. Given the acquisition of the aircraft by the USAF as well as the USN, this was not a surprising total. What was unexpected was the total production that was to follow, for reasons which went back to the beginning of the 1960s.

## Rolls-Royce Phantom

Two events gave the Phantom a second lease of life in the mid-1960s. One was the Vietnam War, which naturally stimulated the US services' need for combat aircraft. The other was the launching of the bi-service TFX programme, which ensured that the Phantom was virtually the only supersonic fighter available to meet that need.

TFX was planned to meet the Navy's requirement for air defence and the Air Force's for an F-105 successor. It was therefore seen as a replacement for the F-4, which the Navy used for air defence and which had replaced the F-105 in near-term Air Force plans. It was also the only new tactical aircraft authorised for either service in the early 1960s, with the single exception of the subsonic Vought A-7. Once production of the F-8 and F-105 was terminated in 1964–65 the F-4 was to be the only supersonic fighter in production for the US services until TFX deliveries started.

This is not the place to discuss the problems that beset the TFX, later known as the General Dynamics/Grumman F-111, but by the end of 1962 the Department of Defense had created a situation in which any slippage or shortfall in the F-111 programme would lead to a large additional requirement for F-4s, even in peacetime. The McDonnell people could not, in their most optimistic daydreams, have foreseen how serious the F-111 problems would be and how many extra F-4s would be needed. Nobody expected that the Navy's TFX would be cancelled completely and that the Navy's next fighter would not even fly until the end of 1970, nor that cost escalation would hack USAF procurement of the aircraft from the planned 1,700 units to just 500. Indeed, the consensus around 1962 would have been that the F-4 line would be reduced to exports by 1965–67, and would probably run out not long afterwards.

Early F-4Cs (top) were externally almost identical to Navy aircraft, down to the colour scheme. The later F-4D (above) was distinguished by a fractionally more bulbous radome.

McDonnell did its best to brighten that somewhat gloomy picture. In February 1962, before the TFX contractor team had been selected, the company made its first proposal for a much improved Phantom, the F4H-2 or F-110B. Heavier than the standard aircraft, it would be powered by a completely different engine: the Allison/Rolls-Royce AR.168-25, developed from the British company's Spey. The AR.168 was a turbofan, combining high thrust in reheat with much improved cruise efficiency. McDonnell claimed that the new Phantom would accomplish 80 per cent of the TFX missions at a fraction of the cost.

Efforts to pre-empt the TFX programme proved unsuccessful but studies of updated F-4s continued. In 1963 Rolls-Royce began to promote the Spey-Phantom for the Royal Navy as a substitute for the Hawker P.1154 (which used an engine from its British rival, Bristol). The RN was unhappy with the P.1154, a radical vertical take-off design which had been originally developed for the RAF, and had admired the Phantom for some time.

The Spey-Phantom took shape as a somewhat different aircraft from the US original. The engines were bigger and needed more air, so the fuselage centre-section and inlets had to be revised. The aircraft would be heavier but would have to land more slowly to fit the smaller British carriers; the wing was therefore modified so that the ailerons would droop more with the flaps, increasing lift on the approach. To maintain control effectiveness in pitch with the ailerons drooped, and hold the nose higher for slow flight, the tailplane was fitted with an inverted leading-edge slat so that it could generate a higher download, and the inboard section of the leading edge was fixed to improve airflow over the tail. The more powerful engines also provided more air for blowing. F-4C-type wheels and tyres were standardized, and the nose

*A Spey-powered YF-4K Phantom FG.1 lifts off from Lambert Field on an early test flight. The bigger nozzles are prominent from this angle.* (MDC)

landing gear was made extensible for deck launches at higher weights.

At about the same time the USAF and USN were looking at Phantom developments beyond the F-4B and F-4D. The two services took different approaches. The USN wanted an improved all-weather interceptor with a new radar; the latter was the Westinghouse AWG-10, the first fighter radar to make use of the Doppler effect to measure the speed and range of its targets and, to some extent, pick up targets against ground or sea clutter. The AWG-10 was more than twice as heavy as the APQ-72, took up nearly twice as much room and required a bigger, bulged radome, but in the Navy's view the improvement was worth it.

The Air Force, with a multi-role fighter/strike mission in mind, was more interested in rectifying the F-4's lack of an internal gun. McDonnell Douglas, Westinghouse and the Air Force worked together to create a Phantom variant with a nose similar in shape to that of the RF-4 and housing a compact "solid-state" derivative of the APQ-72 series, the APQ-120, and a neatly packaged M-61 six-barrel cannon. Again, this meant a weight increase and both the new Navy F-4 and the gun-toting Air Force development emerged with the stronger F-4C landing gear and the control changes proposed for the Spey-Phantom. In turn, the aircraft proposed for Britain acquired the AWG-10 radar. General Electric managed however to defend its US market against Rolls-Royce, and the US services chose an uprated J79 for their new aircraft.

Another new piece of equipment proposed for the new US Navy aircraft combined a two-way air-to-surface data link with an approach power compensator and was intended to permit automatic carrier landings. A small batch of modified F-4Bs, redesignated F-4G, were fitted with this equipment and used for operational trials, and part of the system was used on the new Navy fighter.

All three new derivatives – the British Phantom, the improved Navy fighter and the gun-armed USAF version – went ahead between 1964 and 1966, just as production moved up to a rate of 40 aircraft per month. The Air Force aircraft became the F-4E, the Navy's new aircraft was designated F-4J and the Royal Navy's version became the F-4K. They were joined by the F-4M for the Royal Air Force.

Britain's requirement for the Phantom had become a political issue. In early 1964 a Conservative Party administration shelved the plan to build the P.1154 for both the RAF and RN, leaving the Navy free to negotiate for Phantoms, the order being placed in the summer of 1964. In 1965 the new Labour Government decided to cancel the P.1154 and to buy Phantoms for the RAF as well. But the RAF was not convinced by the Spey-Phantom: they did not need the extra range promised by Rolls-Royce, and the standard J79-powered Phantom was, even then, some 30 per cent cheaper than the Spey version. Then the Government scrapped the Royal Navy's plans for new aircraft carriers, sharply reducing its need for Phantoms. But a decision to abandon the Spey-Phantom would have added one more to a long line of British cancellations in 1964–65 and would have left Rolls-Royce in a critical condition, with virtually no active and profitable engine programmes. On the other hand, there were high hopes that the new aircraft would prove far superior to the original version – shades of the Merlin-powered P-51 – and that it would be attractive to the US services and for export.

This was not to be the case. Even before the aircraft flew, technical problems and cost increases had been encountered. Just before a major defence review Rolls-Royce had to impose a restriction on engine operating temperatures to

Most of the British Phantoms were delivered to the RAF and wore this glossy camouflage finish until the mid-1970s. (MDC)

...ered Phantoms are now mainly used in the air-defence role, armed
Flash missile and the latest AIM-9L Sidewinder. The AWG-11/12
updated to handle the new missile. Air-defence aircraft are being
grey.

equipment for Marine Corps fighter squadrons. Note new ECM enguipment fairings on intake ducts, introduced around 1980. (MDC)

## ng data

ghter

Royce Spey Mk 201 turbofans rated at 12,500lb (5,670kg)
(9,300kg) with afterburning

.7m)
(17.76m)
(4.96m)
m²)

00kg)
(26,300kg)

2.0
ft (15,250m)
ft (1,650km)

## F-4F leading data

**Mission** Air-superiority and interceptor fighter

**Powerplant** Two General Electric J79-GE-17 turbojets rated at 11,870lb (5,385kg) dry thrust and 17,900lb (8,120kg) with afterburning. Cartridge starting

**Armament** normally four Raytheon AIM-7F Sparrow recessed into fuselage, four Ford Aerospace AIM-9L Sidewinder on two dual wing pylons. Internal M61A-1 20mm cannon with 640 rounds

**Avionics** Westinghouse APQ-120 radar fire-control system

**Dimensions**
**Wing span** 38ft 5in (11.7m)
**Length overall** 63ft 0in (19.20m)
**Height overall** 16ft 3in (4.96m)
**Wing area** 530ft² (49.2m²)

**Weights**
**Empty** 31,500lb (14,285kg)
**Max take-off** 61,795lb (28,055kg)

**Performance**
**Maximum speed** Mach 2.2
**Service ceiling** 55,000ft (17,000m)
**Combat radius** 780nm (1,450km)

Visible on the nearer wing leading edge of this F-4E, destined for delivery to the Imperial Iranian Air Force, is the Northrop Tiseo electro-optical telescope for target identification; it was then a new item even for the US services. (MDC)

Packing the M61A-120mm cannon into the same nose profile as that of the RF-4, no easy task, was facilitated by the development of a solid-state (transistorized) version of the Westinghouse radar. The new APQ-120 was less vulnerable to vibration caused by the gun. Extra weight in the nose was balanced by the additional fuel in the tail of the F-4.

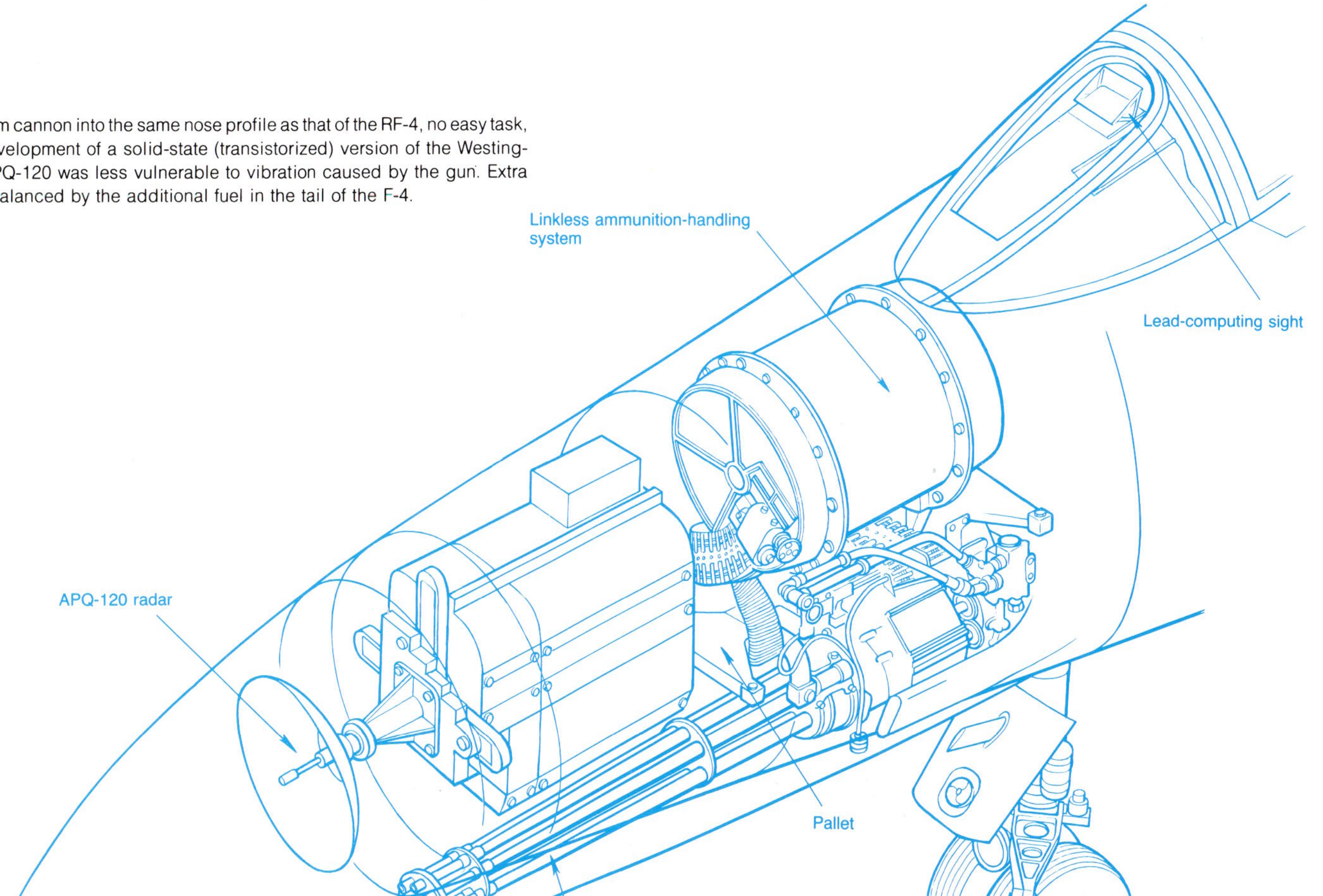

Lead-computing sight

Pallet

Linkless ammunition-handling system

APQ-120 radar

all its own F-4EJs, with the exception of this aircraft and 13 other RF-4EJ craft assembled in St Louis. The ventral pod is an instrument package for (MDC)

**Below** A late-production F-4E, with its leading-edge slats deployed, on a test flight before delivery to the Greek Air Force. Most F-4s exported were of the cannon-armed F-4E model.

James Goulding

GEFAHR
FANGHAKEN
DANGER
ARRESTING HOOK

37+52

contain costs; the snag was that the limitation severely reduced the Phantom's maximum speed. At the same time successive economy measures reduced the total number of aircraft. The Royal Navy requirement was cut from 140 aircraft to 48. Then the RAF requirement was dropped from 150 to 120, though the service did take over half the remaining Navy aircraft. From nearly 300 units the total British Phantom programme had been cut to 170, each aircraft consequently carrying a bigger share of development costs and being proportionately more expensive.

The first F-4K flew in June 1966. Rolls-Royce had already run a Spey behind a full-scale replica of the new intake, which worked perfectly. The reheat system gave serious trouble, however. The flame would sometimes separate from the flameholders, and in some flight conditions the afterburners would "buzz," a phenomenon caused by rapid changes in the flame pattern. The Royal Navy, which had demanded engine handling equivalent to that of the standard aircraft, refused to accept the F-4K until the problems were fixed. This was done successfully and the first aircraft were delivered in April 1968. But the resulting aircraft represented only a marginal improvement over the standard Phantom in some areas – ferry range, for example, was some 15 per cent greater, and the take-off and early acceleration and climb were quicker – and was worse in many others, such as maximum speed. It used more fuel in reheat, had a lower service ceiling and higher drag, and above all was far more expensive. Later, too, it was realised that the RAF could not simply buy extra aircraft to cover attrition because the Spey-Phantom was out of production (the last was delivered in late 1969) and it would have been uneconomical to restart the line.

Apart from the fuselage and intake changes needed to accommodate the larger turbofan engines, the RN F-4K (Phantom FG.1) and the RAF F-4M (Phantom FGR.2) conformed closely to the F-4J, being distinguished by the more bulbous radome for the new-type radar. The F-4K included some special features: its AWG-11 radar was modified to swing sideways to fit the smaller lifts of British carriers, and the nosewheel leg could extend by 40in against the 20in of the F-4J. The two types differed slightly in their avionics suites. The RN aircraft had a Bendix navigation computer while the RAF FGR.2 carried the Ferranti nav-attack system originally designed for the P.1154. The FGR.2 was also modified to carry a large EMI reconnaissance pod – a 22ft-long unit incorporating side-looking radar and infra-red linescan – on the centreline pylon.

Meanwhile, both of the improved US aircraft were proceeding into production. The F-4J was publicly demonstrated on May 27, 1966 – exactly eight years after the XF4H-1 flew – when it was on the point of replacing the F-4B in production. The gun-armed F-4E was flying by that time and was phased into service at Nellis AFB, Nevada, in late 1967. Both aircraft had a new No 7 fuel cell in the rear fuselage, balancing the greater weight which each carried in its nose, and were powered by basically similar J79 engines offering a ten per cent boost in unreheated thrust. The Navy engine was the J79-GE-10 and the USAF powerplant was the GE-17, with cartridge starting and other features.

A US Air Forces Europe F-4E takes a turn as a high-speed target tug.

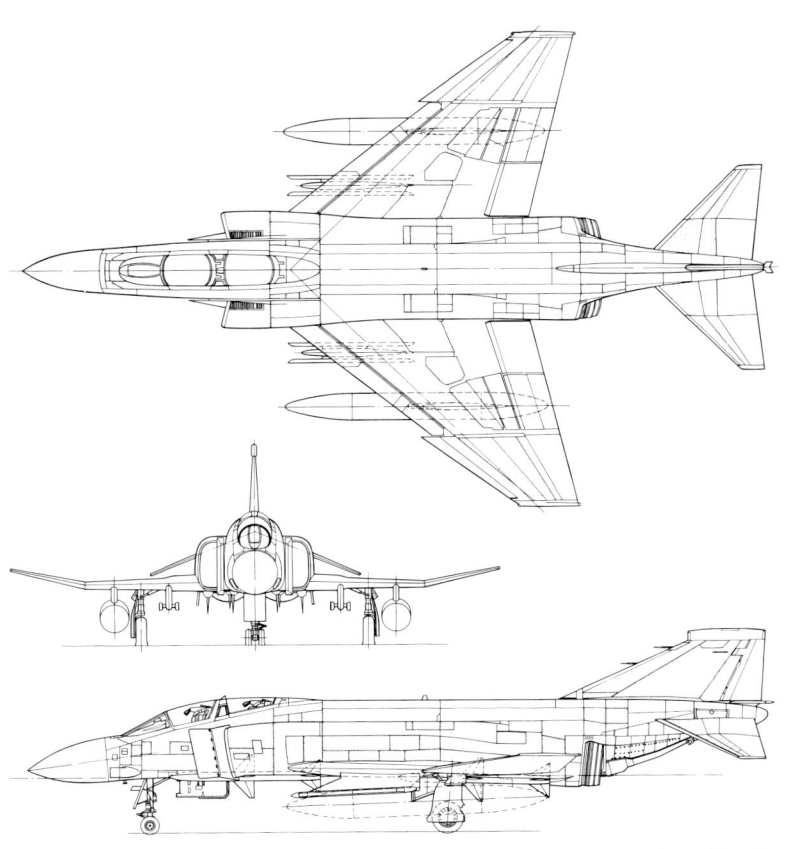

Installation of the bigger Spey engines in the Phantom FGR.2 meant larger inlets, a broader fuselage and wider nozzles. In the 1970s the Phantoms were fitted with standard RAF radar-warning receivers in the fin-top fairing.

The F-4J and F-4E were to represent the final stage in Phantom development for the US services. Neither was a startling advance over its predecessors, but the sum of the various modifications was a notably efficient job of mid-term development. As they entered service it was becoming increasingly apparent that the USAF's F-111 would be too expensive to equip more than a fraction of Tactical Air Command and that the F-111B might not make it into Navy service at all. Meanwhile, as the US poured resources into the Vietnam War and the British order was completed, Phantom production continued to accelerate. In 1967 no fewer than 72 Phantoms were being delivered each month. McDonnell used the profits to acquire control of the ailing Douglas company in that year: the Phantom had created an industry giant.

A napalm-armed F-4C rolls into a bombing run over Vietnam. Sparrows and drop tanks are also carried.

## Trial by combat

Early in 1964 President Johnson ordered the preparation of plans for air strikes against North Vietnam in support of the US-allied government in the South. Up to that point the role of the US Air Force and US Navy air units had been confined to ground support operations by slow attack aircraft and to purely defensive patrol missions in the absence of offensive operations by the enemy. The decision to use air power against the North was to change the picture and would prove to be a difficult and at times traumatic test for the Navy and Air Force. The F-4 played a pivotal role in the war, and more than any other aircraft it was to demonstrate the strengths and weaknesses of the theories of air warfare that had dominated planning since Korea.

A few isolated attacks followed the passing by Congress of the Tonkin Gulf Resolution, which authorized the extension of air strikes into the North. In February 1965, however, Presidential approval was given to a systematic bombing campaign. Code-named Rolling Thunder, the plan comprised attacks on railways, roads and military installations with the aim of preventing North Vietnam from influencing the war in the South. US Navy F-4Bs were based on carriers off South Vietnam from the start of Rolling Thunder, and the first US Air Force F-4Cs arrived in South-east Asia in April 1965.

At the start of Rolling Thunder the F-4s were not heavily involved. The North Vietnamese Air Force (NVAF) consisted of a few dozen MiG-15 and MiG-17 fighters. The MiG-15 had seen action in Korea, where the USAF's F-86 Sabres had shot down ten of them for every Sabre which fell to the MiGs. The MiG-17 was a slightly faster development of the MiG-15. Neither was supersonic and neither was equipped with even heat-seeking missiles. At first the US services saw no need to escort their air strikes in the face of so minor a threat. Moreover, direct attacks on NVAF bases were forbidden by the US forces' own rules of engagement. In neither the Navy nor the Air Force was the F-4 the primary air-to-ground strike aircraft – the USAF made greater use of its F-105s in the strike role – so in the first few months of Rolling Thunder the type saw little action.

Then, in April 1965, NVAF MiG-17s shot down two F-105s during a raid on the Thanh Hoa railway bridge 80 miles south of Hanoi. Clearly the antiquated fighters were a threat to be reckoned with, especially for the bomb-laden strike aircraft. At low altitude and with external stores the F-105s were no faster than the MiG-17s, which operated under tight and efficient ground control. Soon USAF F-4s were being assigned to escort the strike aircraft, while the Navy began to use the Phantom for "MiGCAP," or loose escort of bombing raids. US Navy F-4Bs were the first Phantoms to destroy MiG-17s, in June 1965. USAF F-4Cs were then assigned to MiGCAP, scoring their first kills in the following month.

North Vietnamese air defences continued to improve. In July 1965 an EB-66 electronic warfare aircraft detected radar pulses from an SA-2 Guideline missile system based west of Hanoi. The Soviet Union also supplied improved early-warning and height-finding radar, and in December the NVAF acquired its first supersonic MiG-21 fighters. Rolling Thunder strikes continued in the face of intensifying opposition.

As the NVAF fighters harried the strike aircraft the F-4 pilots would counterattack and try to force them to break off. The result was very often something which the theorists of the 1950s had relegated to history: a close-quarters dogfight in which the objective was to get on the opponent's tail by manoeuvre and by deception. As a form of air warfare it was as far from the thoughts of the Phantom's creators as could be imagined.

Air combat was made more difficult by the rules of engagement that the US forces had imposed on themselves. The most important of these laid down that a target had to be visually identified before it was attacked. This rule alone prevented the Phantom from operating in the way it had been designed to do, since its Westinghouse radar and Sparrow missiles were specifically intended for beyond-visual-range (BVR) interception. Just as important, the designers had envisaged that the crew would perform an interception under radar guidance, flying the aircraft and operating the weapons under the direction of cockpit instruments. The view from the Phantom cockpit was far more restricted than that of the Sabre, MiG-15 or even the MiG-21F, because the canopy had been pared down almost to a minimum to reduce drag and weight.

It was questionable whether BVR interception would have been a workable tactic in Vietnam in any case. Combat quickly proved that the predicted kill probability (PK) figures for US air-to-air missiles (AAMs) were optimistic. It is probably true to say that the Phantom's multi-missile armament saved the USAF from disaster over Vietnam: it was not unknown for an F-4 crew to release six out of eight missiles for a single kill. Some missiles failed to guide, the Sparrow in particular requiring a good radar lock if it was to home on to the target. The AIM-9 Sidewinder was more reliable, though only at short range, and it worked best only if the sun was behind the launching aircraft. Finally, a desperate MiG-17 or MiG-21 pilot could shake off either a Sparrow or a Sidewinder with a well-timed break. The US Air Force F-4Cs arrived in Vietnam equipped with the Hughes AIM-4 Falcon missile – which the USAF had developed while the Navy produced the Sparrow and the Sidewinder – but after disappointing results in early operations the Air Force standardized on the Navy-type weapons.

Despite the Phantom's heavy armament – which threw weight of numbers against the low operational PKs of the individual missiles – the F-4 was not the most successful US aircraft in air-to-air combat over Vietnam. That distinction belongs to the Vought F-8 Crusader. Smaller, less powerful and less well equipped than the Phantom in most respects, the Crusader in 1965–68 had one tremendous advantage: an effective cannon armament. In the first half of 1967 the first SUU-23A Vulcan cannon pods appeared underneath USAF F-4Ds and started scoring immediate kills. But podded cannon were far from an ideal solution. Firing the gun caused the nose to pitch down, and the pod itself was heavy, created a great deal of drag and prevented the carriage of fuel on the centreline pylon. But the Crusader was out of production and suffered an extremely high accident rate, and of course the USAF had no Crusaders. After the F-8 the Phantom was by a very large margin the best aircraft available for air combat.

The F-4's main attributes apart from its weight of fire were its high power and large wing, which gave it the ability to pull tight turns without sacrificing height or speed, known collectively as "energy". At low level and at speeds below Mach 1, where the NVAF fighters operated, this ability was essential. So, on the other hand, was the ability to lose speed rapidly in order to force an opponent to overshoot into a vulnerable position.

Between the start of the Rolling Thunder operations and the end of March 1968, when the US halted bombing of the North, some three-quarters of the USAF strike missions were flown by Republic F-105s. Flying MiGCAP and escort missions, F-4Cs and F-4Ds – the latter being predominant from mid-1967 – accounted for some two-thirds of the kills confirmed by the USAF, destroying about 60 NVAF fighters in air-to-air combat. But the average kill-to-loss ratio in air combat throughout the Rolling Thunder campaign was 2.5:1 in the USAF's favour, only a quarter of the figure achieved by much less expensive aircraft in Korea. Additionally, aircraft were being lost to SA-2 missiles – the weapon scored its first hit on July 24, 1965, destroying one F-4C and damaging three others in the same formation – and to the intensifying anti-aircraft artillery (AAA) defences.

On the other hand, some operations during Rolling Thunder showed that the Phantom was by no means an inferior machine provided that it could be freed from some of the operational restrictions imposed by national policy. In January 1967, for example, the 8th Tactical Fighter Wing led by Col Robin Olds conducted Operation Bolo, the largest all-fighter operation of the war. The purpose of Bolo was to simulate a normal bombing raid and draw the defending fighters into a large formation of F-4s. The 14 four-aircraft flights succeeded in destroying seven MiGs without loss. But by June 1968 the USAF was losing more aircraft than it was shooting down, a development which was later linked to the influx of less experienced pilots.

The Air Force, however, did not immediately respond to the problem by re-introducing air combat manoeuvring (ACM) into its training programmes. The Navy had come to a similar conclusion – essentially, that the F-4s were not dominating their smaller and less sophisticated opponents as they should have done – but responded in a different manner. In March 1969, almost a year after the temporary suspension of the air war over Vietnam, the first class of a unique academy commenced at NAS Miramar. This was Top Gun, formed as a result of an analysis of the Navy's air-to-air combat record.

The five-week Top Gun course was intended to turn the most promising pilots from individual squadrons into ACM experts who would pass on their knowledge to the rest of the unit. At the same time the USN increased the emphasis on ACM in regular squadron training, despite an increase in the rate of training accidents. In the process the Navy pilots learned to cope with some of the less attractive characteristics of the Phantom, which had become serious problems in the

A Vietnam-veteran F-4C, victor of two battles with enemy fighters, in post-war service with the Louisiana Air National Guard. (MDC)

world of ACM. The worst of these was the stall/spin behaviour of the F-4. If the wing was stalled the aircraft would very quickly snap into a rapid and vicious spin from which recovery was difficult at any altitude. At the low levels characteristic of air combat over Vietnam these "departures" often ended in a crash. Another characteristic was "Mach tuck": the Phantom's stabilator became more effective at subsonic speeds, so a transonic deceleration could make a 2g pull-up increase abruptly to 5g or 6g as the stabilator bit into the air. A mid-1970s commander of Top Gun said: "To be successful in an F-4 [in air combat] you have to fly outside the envelope – you have to be able to depart it and recover it." In short, the Navy was teaching its pilots to fly the F-4 beyond all the limits and then recover.

Another pioneering aspect of Navy training was dissimilar air combat manoeuvring (DACM), or mock combat with different types of aircraft. Navy F-4 pilots flew against A-4 Skyhawks and F-106s, types which in some respects had similar combat characteristics to the NVAF MiGs. This part of Top Gun also got under way in 1968–69.

From 1968 to 1972 the total of US forces in Vietnam fell from 540,000 men to 70,000 and direct air operations against the North ceased completely. Air combats were rare, USN kills slowing to a trickle and USAF kills stopping completely. By the end of March 1972 only 64 USAF Phantoms were available in Vietnam, with another 150-plus in Thailand. Combat operations had been confined to reconnaissance and a few special missions: for example, the 497th TFS, based at Ubon in Thailand, was the only nocturnal Phantom unit and took part in the interdiction of the Ho Chi Minh Trail through Laos.

In this time of uneasy truce, as the US Government tried to build up the South Vietnamese system to withstand the determined forces that would destroy it, and the Navy practised air combat, the Air Force concentrated on improving its capabilities in the strike role. The basic problem in Rolling Thunder had been that the vast majority of the weapons used simply missed. The Bullpup ASM had proved almost completely ineffective, while conventional "iron bombs" lacked the accuracy to hit precision targets, even when dropped in hazardous low-level straight runs.

The Navy had pioneered the use of a better mousetrap in the shape of Walleye, which had been under development before the Vietnam War started and had been introduced in January 1967, though it had not been used on F-4s. Walleye was a winged, unpowered weapon. Its nose contained a television camera which produced a high-contrast picture of the view ahead. The pilot aimed the camera on to the target and released the Walleye, and the weapon's own internal circuitry would guide the weapon on to the televised target.

By 1972 the US Air Force had deployed its own version of this system in the shape of the Paveway family of weapons. These were standard heavy bombs – 2,000lb and 3,000lb weapons – fitted with stabilizing and lift fins and alternative "steering modules" consisting of movable fins and a guidance system. The latter was either TV-based, like the Walleye system, the weapon being referred to as an EOGB (electro-optical guided bomb), or designed to home on the reflection from a laser beam directed on to the target, in which case the weapon was termed an LGB (laser-guided bomb). The carrier aircraft was fitted with an Aeronutronic AVQ-10 Pave Knife pod which combined television acquisition equipment and a laser "designator".

Another important change had taken place in the USAF by 1972: the extensive use of the F-105 in the Rolling Thunder operations had cost the service so many aircraft destroyed or damaged beyond repair that the type had to be confined to special missions. From that point onwards the primary USAF strike aircraft was the Phantom.

The NVAF defences had been considerably reinforced during the bombing halt. It was estimated that by the first half of 1972 the NVAF had 250 fighters, of which one-third were second-generation, improved models of the supersonic MiG-21. More light, agile MiG-19s were also in service. Many of the remaining MiG-17s had been equipped to carry

36

AAMs. Some 300 SA-2 sites were operational, backed up by 1,500 AAA installations.

The ceasefire broke on 30 March, 1972, when General Giap's invading forces rolled into the South. Four more USN carriers were directed to the Vietnam theatre, while Tactical Air Command launched the biggest series of deployments in its history under the code name Constant Guard. One of the main objectives was to get as many as possible of the new F-4Es to Vietnam in case the NVAF's fighters were deployed in defence of the invading forces. Most of the F-4Es were based in the USA, 8,000 miles from the war. Constant Guard I started on April 8, with 18 F-4Es of the 334th TFS being transferred from Seymour Johnson AFB, North Carolina, via Hickam and Guam to Thailand: three sectors with multiple air refuelling on each, flown in cockpits made even more cramped by bulky survival kit. Another 18-aircraft F-4E squadron followed 24hr later. Constant Guard II, in early May, moved a similar number of aircraft, while the biggest single operation was Constant Guard III, under which the entire 49th Tactical Fighter Wing – four squadrons with a total of 72 F-4Ds – transferred from Holloman AFB in New Mexico to Takhli in Thailand. Only one F-4 dropped out en route. By the end of May there were 400 "fast jet" strike aircraft in the theatre, the majority of them F-4Ds and F-4Es. The aircraft were equipped with Paveway laser-guided bombs and improved versions of the Sparrow and Sidewinder.

At the beginning of May 1972 this force was ordered into a new bombing campaign, entitled Linebacker (later known as Linebacker I), against North Vietnam. From the very beginning Linebacker I differed from previous operations in that individual units were assigned to different tasks. Most units worked on developing the skills needed to deliver LGBs and EOGBs efficiently, while others concentrated on air-to-air combat or defence suppression.

A typical Linebacker I mission could involve 50–70 aircraft, of which as few as 32 might be carrying weapons to the primary target. These "strike package" aircraft would be carrying LGBs and EOGBs, weapons which quickly proved their worth. Between April and June 1972 the F-4s of the 8th TFW destroyed no fewer than 106 bridges with the new weapons, including some which had resisted repeated and costly strikes in the early phase of the war. Two of the most difficult targets of the pre-1968 era, the Paul Doumer and Thanh Hoa bridges, went down within days of the start of Linebacker I. Because of the vast improvement in the accuracy of the new "smart bombs" far fewer of them needed to be delivered to score hits, allowing the Air Force to economize on both bombs and aircraft. While a pre-1968 "iron bomb" F-105 might have carried up to a dozen 750lb bombs, a Linebacker I F-4 might have only a pair of 3,000lb LGBs.

The only problem with the smart bombs was that their delivery involved a long straight run at moderate altitude – the LGB was accurate from as much as 14,000ft – and often in close formation. The more accurate LGB had to be lobbed into a "basket," a cone-shaped volume of sky within which the weapon could pick up the target designated by the leader, and to do this the attackers had to keep a four-span separation distance. (The EOGB was generally used when weather conditions were hazy enough to make the LGB inaccurate.) The necessary close formation could make the entire flight vulnerable to a MiG or SA-2 attack, so while the number of strike aircraft required for a given mission decreased, the job of escorting them became more demand-

A Thai-based F-4E drops "fuse-extended" Mk 82 bombs from medium altitude over North Vietnam. (MDC)

ing. It was for this reason that the support element of a Linebacker mission could number as many as 40 aircraft.

Some of these support aircraft would be direct escort aircraft. Others would be assigned to the MiGCAP role, still supporting the strike force but more loosely attached to it. F-4s in Linebacker operations also began to fill more specialized roles, and 1972 saw the first use of the Phantom in defence suppression. Some of these were Iron Hand strike aircraft, intended to hit SAM and AAA sites with cluster weapons; these shotgun-effect bombs were highly effective against thin-skinned targets. Other Phantoms began to be used in a still more demanding and important mission. These were Wild Weasels, equipped to detect and destroy SAM sites or, at least, to intimidate them into shutting down their radars. The Weasel was a two-seat fighter carrying a radar homing and warning system managed by an electronic warfare operator (EWO) in the back seat, and armed with cluster bombs and, later, AGM-45 Shrike missiles which homed on to radar emissions. They were so named because they were a more aggressive extension of electronic intelligence or "ferret" operations; the back-seat EWOs became known as "bears". The first Wild Weasels were F-100Fs, though most of the Vietnam Weasels were F-105Gs. But because the supply of two-seat Thunderchiefs was strictly limited they were joined by a few F-4Cs of the 67th TFS; these were known as EF-4Cs and were operational in 1972.

The Weasels and Iron Hand aircraft were the vanguard of the strike formation. Two to three minutes behind there followed a formation of between four and eight F-4s or A-7s laden with bulky pods full of aluminium-foil strips or "chaff". The oldest electronic countermeasures (ECM) device, chaff created powerful echoes on radar screens and masked the strike force from SAMs. The chaffing aircraft were however vulnerable because they needed to maintain a steady course and direction. They were protected from SAMs by podded radar jammers which caused the missile radars to confuse different aircraft in the formation, and they were accompanied by escorts to keep MiGs from jumping their tails and forcing them to jettison their vital loads prematurely.

Behind the chaffers came the strike aircraft and their escorts, while the MiGCAP roved above the ingress route and the combat area. All elements used high airspeed as their main protection; the USAF had not yet started the intensive programme of air combat training that the Navy had used. Despite the great improvements embodied in the new Sparrow and Sidewinder missiles, and the introduction of the F-4E with its improved weapon-aiming systems and internal gun, the USAF's air combat kill/loss ratio remained below 2:1 during Linebacker I, with another 17 aircraft lost to SAMs and AAA. Another notable feature of the last year of the war was that all the USAF kills were claimed by F-4 crews apart from two MiG-21s downed by B-52 gunners during Linebacker II, the final air offensive unleashed in December 1972.

Combined with the success enjoyed by Navy crews such as Lts Randy Cunningham and Willie Driscoll, the first aces (five confirmed kills) of the Vietnam War, the Linebacker I results persuaded the USAF to re-emphasize air combat training. But the Air Force was already committed to a programme which would markedly improve the combat capability of the Phantom when the Linebacker I strikes started, though the idea did not bear full fruit until the US air operations in Vietnam had ceased.

The USAF effort was aimed at curing the Phantom's stall/spin problem, which by late 1970 had claimed at least 80 aircraft destroyed, significantly more than the NVAF score at that time. In early 1971 the Department of Defense ordered a research programme, in response to which McDonnell Douglas proposed a radical modification to be applied to existing and future F-4Es. The wings were to be strengthened and the existing blown leading-edge flaps removed. They would be replaced by powered slats, automatically activated by the flight control system, which would extend at high angles of attack, simultaneously delaying the stall and warning the pilot that a final stall was approaching. Not only did the modifications extend the F-4's manoeuvring envelope, but their warning function also allowed the pilot to approach the edges of the envelope less nervously. The snag was that the slats could not be blown like plain flaps, so to keep approach speeds the same as those of the standard aircraft the leading-edge shape had to be biased more towards the low end of the speed range. This blunter leading edge created much more drag at high speed, with the result that the slatted F-4 was markedly slower than the standard aircraft. But the maximum speed of the F-4 had seldom if ever been used in Vietnam, and the improvements in manoeuvrability were judged well worth the loss in speed.

The last US Phantom kill was on January 8, 1973, on a night MiGCAP mission. It was two years later that the USAF F-4 next fired weapons in anger, possibly for the last time, when Phantoms were used to attack Cambodian posi-

The single most numerous version of the Phantom was the F-4E, the only model with built-in cannon armament. Most F-4Es were later fitted with automatic leading-edge slats, as seen here, and strengthened wings.

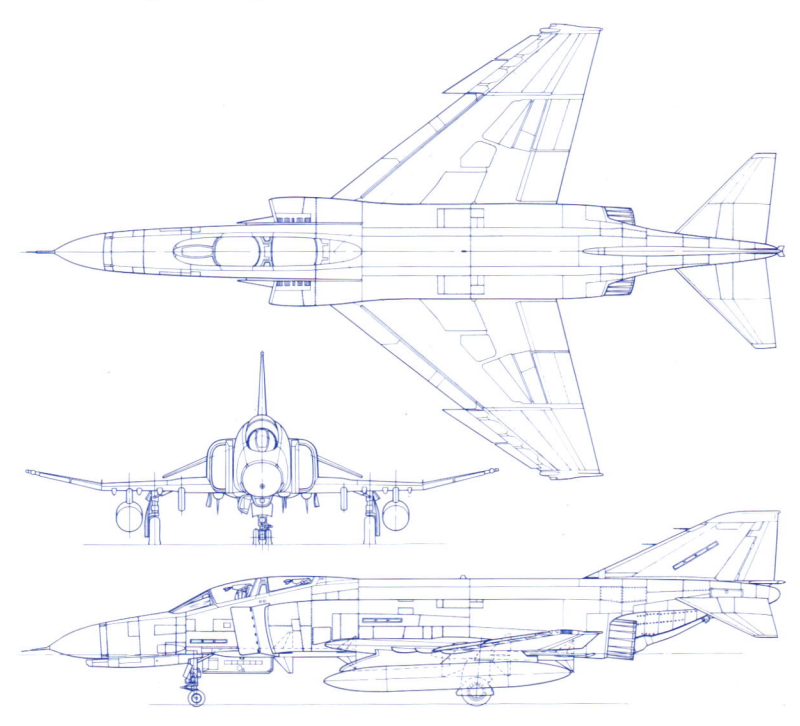

Three F-4Ns (rebuilt F-4Bs) from the USS *Coral Sea* formate over the California coast. (MDC)

tions during the attempt to free the crew of the SS *Mayaguez*.

Vietnam gave the Phantom an assured place in history. The performance of the aircraft in combat, in both its positive and negative aspects, was central to the lessons which the USAF, USN and other services learned from the conflict. Pressed into roles for which it had never been designed, the Phantom nevertheless gave a far better account of itself in air combat than could most of its contemporaries. But this use of the aircraft pointed out the need for better manoeuvrability, still greater acceleration and improvements in the crew station. The Phantom's contribution to aviation history lies in the fact that, having been designed according to one theory of air warfare, it was flexible enough to become the flying laboratory for new and better ideas when that theory proved unsound.

## *Updates and exports*

By the end of the US forces' involvement in Vietnam the F-4 had already gone into battle under other flags, was flying with the air forces of half a dozen nations and had been ordered by a few more. Even at well over twice the price of an F-5 or Mirage, the Phantom proved attractive to export customers. Had the richest of all export markets not been pre-empted by the F-104 while the Phantom was still an unknown quantity, it would probably have sold even more widely.

The biggest user of the Phantom remains the United States, and while production of the type for the US forces had slowed to a trickle by the mid-1970s it is still an important item in the inventory, for the same reason as accounted for its importance in Vietnam: its replacements have been expensive and, in some cases, slow to appear.

The US Navy finally buried the F-111B in 1968, quickly substituting the Grumman F-14 Tomcat, designed around the same radar/weapon system and engines. The Tomcat was developed with commendable speed, being deployed aboard USS *Enterprise* in September 1974. The snag was that the unit price had climbed above $20 million, almost three times the price of an F-4J. (Production of the F-4J had ended in 1972.) McDonnell tried unsuccessfully to unseat the F-14, first with an improved Phantom powered, like the British aircraft, with turbofan engines, and then with a version of the F-15. Instead, the Navy opted to avoid the problem of the F-14's cost by developing a smaller and cheaper fighter/attack type to replace some of its F-4s and other attack aircraft, so that fewer of the expensive Tomcats would be needed. But this meant another complete procurement process, from design competition through development, production and service introduction. The resulting aircraft, the McDonnell Douglas F/A-18A Hornet, was delivered to test units in the second half of 1981. Meanwhile, the F-4s have soldiered on to keep Navy and Marine squadron strength up.

Even some of the F-4Bs – ten-year-old aircraft by the mid-1970s – have been retained in service. Some 228 aircraft have been through exhaustive inspection and repair, and have been modified with the stronger basic wing – without the slats – of the F-4E. Used by Marine and reserve units, these aircraft are designed F-4N.

In the early 1970s the F-4J force was retrofitted with the improved AWG-10A radar, including a digital computer, built-in test equipment and an improved optical sight. The decision to acquire a second new type of fighter led however to a need for a more comprehensive improvement and life-

39

extension programme. The Navy had not taken part in the original slat programme, relying on simplified stall-recovery procedures, but decided to adopt the modification for the improved J. About 300 F-4Js were modified to F-4S configuration, with slatted, strengthened wings and other modifications to increase permissible landing weights and yield an eight-year life extension. These aircraft are likely to be in service with front-line fighter units until the mid-1980s. Most of the Marines' remaining RF-4Bs – 30 aircraft out of a total of 46 delivered – have been rebuilt and equipped with a new inertial navigation system. The F-4S, with slats and new weapons, particularly the formidable AIM-9L Sidewinder, is still a highly useful interceptor, even in the 1980s.

Some of the older F-4s have been assigned to a less glamorous role, acting as targets for the USN's advanced weapon systems, particularly air-to-air missiles. The QF-4B conversion was produced by the Naval Air Development Center at Warminster, Pennsylvania, and the first aircraft was delivered to the Pacific Missile Test Center at Point Mugu in early 1972. Seven QF-4Bs had been converted by 1981. The aircraft are equipped with remote piloting gear, photographic and electronic scoring systems and ECM, and are used for live, life-size simulation of high-g aircraft targets. A more advanced QF-4 was under development in 1982. The US Air Force also plans to use F-4 drones to replace the QF-100 Super Sabre, but as yet there are insufficient spare F-4s to meet the USAF's very large requirements for drones.

The US Air Force's new-aircraft programme has proceeded more rapidly than the Navy's. Like the USN, the service started work on a powerful twin-engined fighter in the late 1960s but found that the resulting aircraft – the F-15 Eagle – was too expensive to counter the more numerous Soviet opposition within a realistic budget. However, by the later years of the 1970s Tactical Air Command had two aircraft in production which were cheaper than the F-15. These types – the A-10 Thunderbolt and F-16 Fighting Falcon – have hastened the retirement of the older F-4D from TAC, US Air Forces Europe and Pacific Air Forces units. The F-4 nevertheless remained the most numerous type in service throughout the 1970s, and the later F-4E has been the subject of several retrofit and update programmes.

One of the first new systems to be fitted to the F-4E

VF-121, the first Phantom unit, finally retired its F-4s in 1980. This farewell formation comprises two F-4S – note slats on nearest aircraft – and (centre) an F-4J. (MDC)

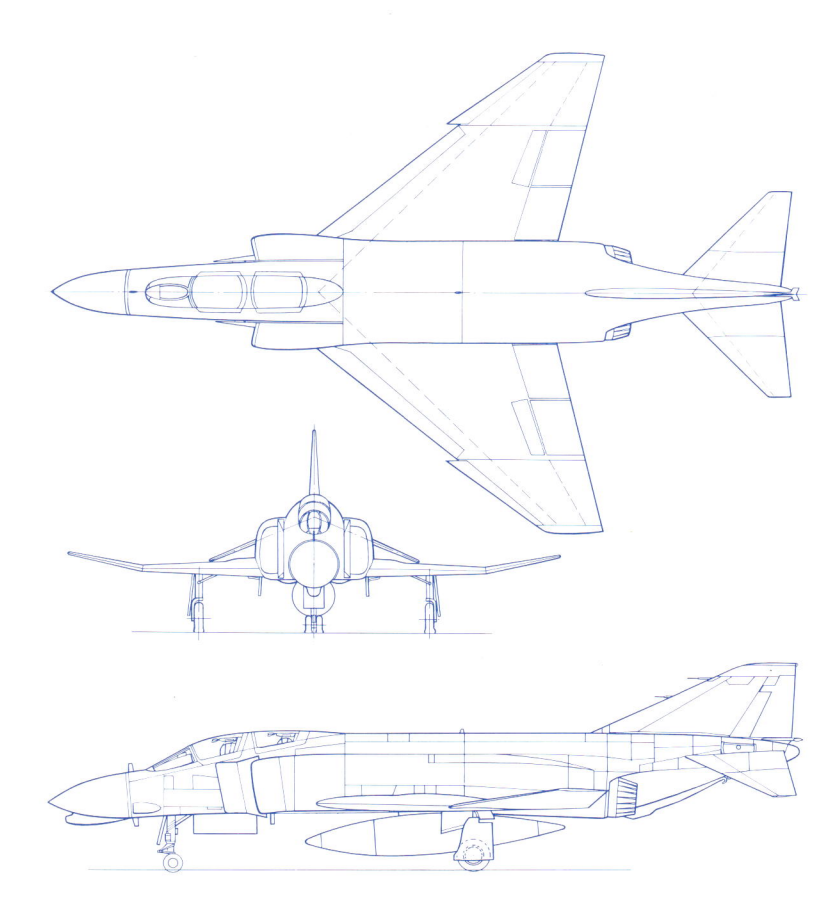

The QF-4B, a stripped-out B fitted with remote piloting gear and electronic scoring equipment, is used for realistic tests of air-to-air weapons against manoeuvring fighter-sized targets. More F-4s will probably be converted to targets as combat units release them.

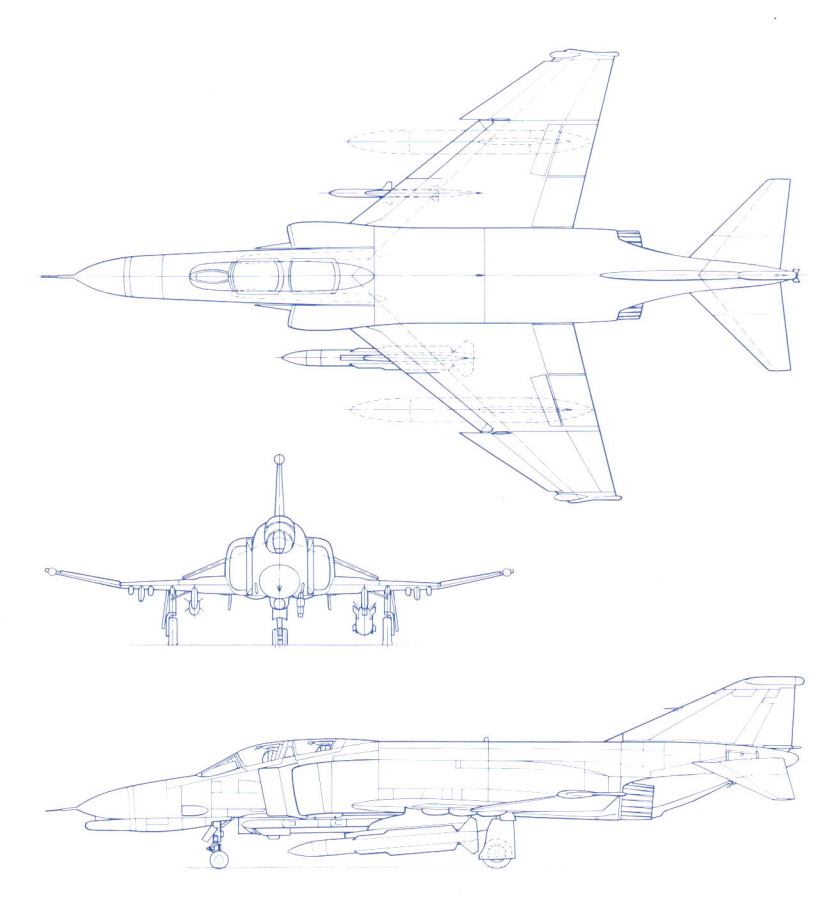

Designed to seek out, menace and destroy opposing surface-to-air missile systems, the F-4G Wild Weasel is a modified F-4E with advanced electronics and an armament of anti-radar missiles. One of the Sparrow bays is occupied by an ALQ-119 ECM pod.

was the Northrop Tiseo (Target Identification System, Electro-Optical), designed to ease the pilot's task in environments like Vietnam in which positive visual identification is required before an attack. Tiseo consists of a television camera mounted on the port wing leading edge, and gives the WSO a telephoto-TV view of the target for easy identification. More than 500 F-4Es were fitted with Tiseo in the mid-1970s.

A later programme covered the installation of the Lear Siegler ARN-101 digital navigation and weapon-aiming system in more than 250 F-4Es and RF-4Cs. From August 1980 some of these aircraft also had a measure of night/all-weather strike capability, carrying the large and sophisticated Ford Aerospace AAQ-9 Pave Tack sensor pod. Pave Tack is a completely self-contained unit incorporating a gimballed and stabilized platform to cover the complete hemisphere below the aircraft. The platform carries forward-looking infra-red (FLIR) equipment for target acquisition, plus a laser for target ranging and designation. The Pave Knife daytime laser/EO pod, used in conjunction with Paveway bombs in Vietnam, has been replaced by the cheaper ASQ-153 Pave Spike from Westinghouse.

The APQ-120 radar of the F-4E has been updated to handle the improved AIM-7F Sparrow missile, and has been modified with more reliable and efficient digital components, including a new missile-release computer. Another new weapon, employed to a limited extent in Vietnam, was the Hughes AGM-65 Maverick, a stand-off ASM with EO guidance.

The biggest single USAF modification programme, however, has seen the Phantom finally replace the F-105 in the Wild Weasel role. After the F-4 had been selected as the new Wild Weasel platform, 116 of the USAF's newest F-4Es were designated for conversion. Initially referred to as the EF-4E, the new version is now the F-4G, re-using the designation applied to a small batch of modified F-4Bs in 1965. The first of the new F-4Gs was delivered from the conversion line at Ogden Air Logistic Center, Utah, in April 1978, and after some development problems the first unit was formed at George AFB, California, in March 1981.

The F-4G has no internal gun, and is modified with the smokeless J79-GE-17C engine and provision for a 600 US gal ventral fuel tank. But the biggest element in the $2.8 million conversion is the McDonnell Douglas APR-38 system: under the control of the "bear" in the back seat the APR-38 detects and locates potentially hostile radar systems and controls the release of the F-4G's weapons. No fewer than 52 antennae, including forward and side-looking interferometer arrays which fill a chin pod where the cannon used to be, pick up emissions which are then identified and classified by means of a library of known systems in the Texas Instruments processor. The system then displays the type and position of threats on CRT screens in both cockpits, assigning priorities to the top 15 threats and marking the worst single danger with a triangle.

The F-4G's armament includes AGM-45 Shrike and the bigger AGM-78 Standard anti-radar missiles, and it is also expected to carry the AGM-88 Harm (High-speed ARM). Maverick can also be carried for close-in attacks on non-emitting targets. An ALQ-119 ECM pod, for defensive elec-

**Above** The nose electronics pod, extra antennae and WW tail code identify this Phantom as an F-4G Wild Weasel defence-suppression aircraft. Some 116 F-4Gs were produced by converting late-production F-4Es. (MDC)

**Below** RF-4Cs in USAF service are being updated with new optical and electronic sensors, improved electronic surveillance measures and new data links. At present there are no plans to retire or replace the aircraft, the last of which was delivered in December 1973. (Don Lunn)

Most of the specialized equipment on the F-4G is carried in the space left by the removal of the gun and ammunition feed. The weapons shown here – short-range Maverick, Vietnam-era Shrike and heavyweight Standard ARM – are to be supplemented by the new Texas Instruments AGM-88 by 1985.

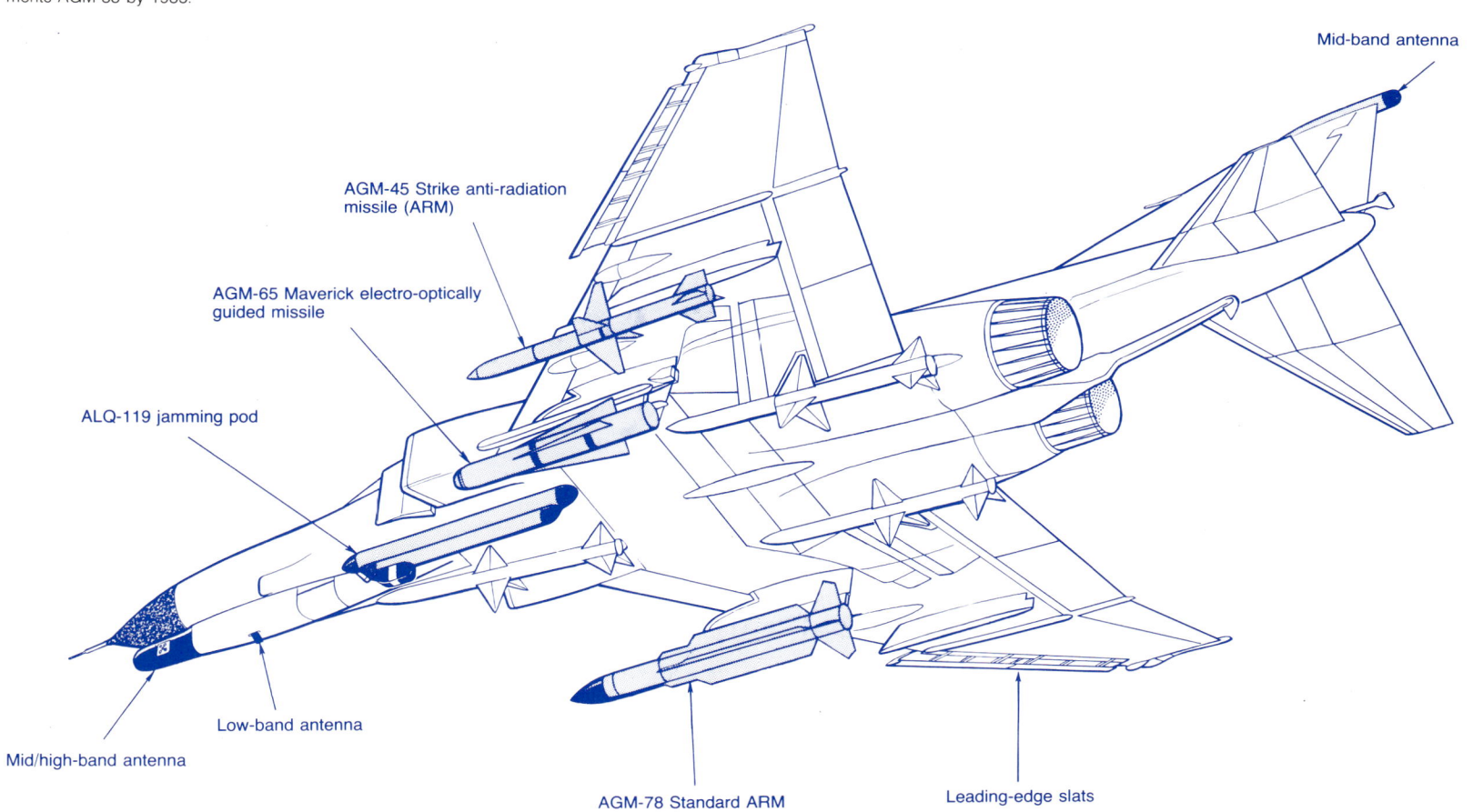

- Mid-band antenna
- AGM-45 Strike anti-radiation missile (ARM)
- AGM-65 Maverick electro-optically guided missile
- ALQ-119 jamming pod
- Mid/high-band antenna
- Low-band antenna
- AGM-78 Standard ARM
- Leading-edge slats

The USAF Thunderbirds equipped with F-4Es in 1969. The fin of the aircraft nearest the camera is blackened by smoke because it occupies the rear slot in many formations. (MDC)

tronic coverage, occupies one of the Sparrow launchers, chaff dispensers are built into the pylons and the F-4G is also designed to carry AIM-7 and AIM-9 missiles for self-defence. Weasel pilots are among the most experienced in TAC, highly proficient in air combat and flight-lead qualified; analysis of Red Flag exercises at Nellis AFB has shown that Weasels are engaged more often than any other aircraft. The F-4G Weasel is expected to be viable against Soviet threats into the 1990s, and is thus likely to be the last front-line combat version of the Phantom.

Also continuing in USAF service, with no sign of a complete replacement is the often neglected RF-4C, which remained in production alongside the later fighter versions. The RF-4Cs and older RF-101s bore the brunt of the highly dangerous reconnaissance missions over Vietnam, following every strike to assess damage to the target, and RF-4Cs were the only USAF tactical reconnaissance aircraft in service up to the end of 1982. Some of them are being displaced by Lockheed TR-1 high-altitude reconnaissance aircraft, but at $40 million a unit the new Lockheed will not meet all the service's needs. Lockheed's F-19 reconnaissance/strike Stealth aircraft may fill the gap to some extent in the second half of the 1980s, but as yet there is no direct replacement for the RF-4C in sight, although versions of the F-15 and the arrow-wing F-16XL have been proposed. The unslatted RF-4Cs, incidentally, are probably the fastest F-4s in service at present.

The Phantom has gained one other unique distinction during its US career: it was the first and so far the only aircraft to be flown by both the US Air Force and US Navy demonstration teams. The Navy's Blue Angels received the first of seven F-4Js – drawn from a batch of trials aircraft – in

Like the Thunderbirds, the US Navy Blue Angels converted to the F-4 in 1969, using F-4Js drawn from the trials batch. (MDC)

late 1968 and gave their first F-4 show in March 1969. The USAF Thunderbirds followed suit with their new F-4Es three months later. While the Phantom was short on grace, there was no denying that formation aerobatics with twenty-ton Mach 2 fighters provided plenty of spectacle and unparalleled exercise for the eardrums. But the F-4 at low altitude and high speed is an out-and-out gas-hog, and in the aftermath of the 1973 oil embargo, domestic fuel price rises and petrol shortages made the continued use of the F-4 politically inadvisable, obliging both teams to re-equip with smaller aircraft.

The F-4 remains in service with nearly all its original export operators. As noted above, the first of these was Britain, which placed its first order in 1964 and took delivery of the last of 170 aircraft in late 1969. The Royal Navy's Phantom FG.1s served throughout the 1970s on HMS *Ark Royal* and were transferred to the RAF when the ship was taken out of service in 1978. The RAF Phantoms entered service at about the same time, almost all of them in the strike/reconnaissance role. The exception was the squadron of Navy-standard FG.1s which were delivered directly to the Air Force and used for interception. As Jaguars were delivered to Strike Command and RAF Germany the Phantom FGR.2s were refurbished after their gruelling low-level careers and re-assigned to air defence in Germany and the UK. Late in 1977 a major force improvement took place when the Phantoms were equipped with the British Aerospace Sky Flash missile, a British development of the Sparrow with a much improved homing head and fuze. Sky Flash is claimed to fall between the latest US Sparrow developments and the all-new AIM-120 Amraam in terms of reliability and

Not one, but three Israel Defence Forces F-4Es. These aircraft have been continually updated and may be re-engined with Pratt & Whitney's new PW1120. (MDC)

lethality. The Phantoms will probably be replaced by Panavia Tornado F.2s in the mid to late 1980s.

Unexpectedly, more Phantoms were acquired by the RAF in the course of 1983. About a dozen ex-USN F-4Js were supplied to RAF squadrons in Germany, replacing aircraft which had been deployed to the South Atlantic following the Falklands War in 1982.

The second customer to announce an F-4 order was Iran, which contracted for two squadrons of F-4Ds in late 1966; these were the last F-4Ds to be completed and had the F-4E-type wing. The first four aircraft were handed over to the 306th Fighter Squadron of the Imperial Iranian Air Force in September 1968, and the last was delivered in November of the next year. Later Iran ordered 177 F-4Es and 16 RF-4Es, all of which had been delivered well before the revolution of late 1979. The renamed Iranian Islamic Air Force has made some use of the aircraft in the Iran-Iraq war, but the supply of spares has become an increasing problem and few of the aircraft are still operational.

The next member of the Phantom club was Korea, which took delivery of its first F-4Ds – supplied from USAF stocks – in August 1969. Some 18 F-4Ds were followed by a

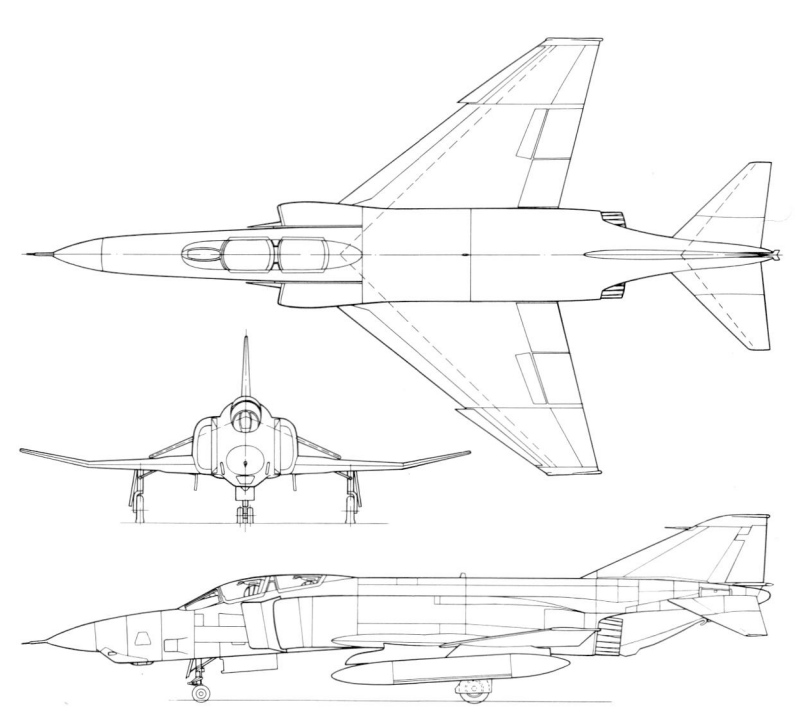

The RF-4E, first adopted by Germany and Israel, combines the RF-4C nose with the more powerful engines and other improvements of the later F-4E. Operational equipment standards differ from customer to customer.

46

similar quantity of F-4Es, and another batch of F-4Es has been supplied recently.

At the end of 1968 Phantom exports took off. Within a space of little more than two months, two large and well equipped air forces selected the Phantom and a third was authorized by the US Government to receive the aircraft. The last-mentioned country, Israel, was the first to take delivery of its aircraft, being followed first by Germany and then by Japan.

The supply of F-4Es to Israel followed an urgent request. After the Six-Day War in June 1967 France had refused to deliver 50 Mirage 5J supersonic strike fighters which had been ordered by Israel, and an initial batch of 50 F-4Es and six RF-4Es was ordered to fill the gap left by the Mirages. Following pilot training by the USAF, the first aircraft became operational in late 1969 and were almost immediately used in action in the War of Attrition against Egypt, carrying out ambitious and dangerous long-range strikes against targets in the Nile Valley and around Cairo. These raids encountered steadily stiffening opposition from Soviet-supplied and Soviet-manned missile systems – including the formidable new SA-6 Gainful, which had not been used in Vietnam – and had to be suspended. Hostilities continued, however, and F-4 losses were heavy until the US managed to bring about a ceasefire in August 1970. Before that, on July 25, Phantoms and Mirages had taken on Soviet-piloted MiG-21s in a 30-aircraft dogfight in which the Israelis shot down five of their adversaries without loss. Meanwhile, the Israelis had tried to cut off Port Said, on the Suez Canal, from the rest of Egypt by air interdiction attacks against bridges, roads and water pipelines connecting the city to the interior. Soviet-manned missile defences countered with massive barrage attacks, inflicting serious F-4 losses. Numbers were restored after the ceasefire by continued deliveries from the USA, including some aircraft from McDonnell Douglas and, later, replacements drawn from US Air Force stocks.

Three years later Israel's Phantoms were in combat again, leading the Israeli counter-attack against the Arab invasion on October 6, 1973. Once again, losses of Phantoms and A-4 Skyhawks were heavy, and again the main danger was from SAMs and AAA. No single system was particularly dangerous: the problems were that the Israelis (and, at that time, the USAF) had no effective jamming systems to use against the SA-6 Gainful or the lethal Shilka with its quadruple 23mm cannon, and that the missiles and guns in service provided a solid block of anti-aircraft coverage down to ground level over any target of value. As most of the sorties needed were ground support operations, a great many Phantoms were lost. The Phantoms carried chaff to confuse radars and flares to decoy infra-red homing missiles such as the SA-7 Grail, and employed combinations of fast, level run-ins and steep diving attacks to take advantage of the main weakness of the SA-6, its slow elevation and depression speed. Before the end of the war replacement aircraft and new electronic-warfare pods were delivered in an emergency support airlift from the USA, and losses declined from the peaks of the first few days. In air-to-air combat – mostly on defensive operations, the Arabs relying on missiles for their own defences – the Israeli Air Force achieved a high kill/loss ratio.

Post-1973 reinforcements and replacements brought total F-4 deliveries to Israel to 204 F-4Es and 12 RF-4Es by the end of 1978, most of the aircraft being transferred from USAF stocks. In the late 1970s these aircraft equipped seven squadrons, indicating that some 120 aircraft plus reserves remained in service after the losses of the War of Attrition and the Yom Kippur War.

Between 1971 and 1973 Israeli Phantom crews made several determined but unsuccessful attempts to intercept MiG-25 Foxbat-B reconnaissance aircraft which intruded into Israeli-held airspace on sorties from Cairo West. The APQ-120/AIM-7E combination was not however equal to snap-up attacks on 1,600kt, 75,000ft targets, although the Israeli pilots are said to have come quite close to their targets with well timed zoom climbs. In part it was the frustration of watching Foxbats sail insolently over their defences that led the Israelis to acquire McDonnell Douglas F-15s, which took over as the top-line Israeli fighter in late 1976. Also delivered in the late 1970s were the first of Israel's F-16s. However, the F-4s were still used extensively for ground-attack purposes in the 1982 invasion of Lebanon, and with the help of unmanned decoy aircraft and sophisticated electronics they avenged themselves on the SA-6. All Israel's Phantoms have been fitted with slats and smokeless engines, and a more sophisticated multi-mode radar produced by Israel's Elta Electronics has been fitted in place of the APQ-120. This sensor feeds a new digital/inertial nav-attack system, making Israel's Phantoms the best equipped anywhere. They will probably remain in first-line service until they are replaced by the new IAI Lavi fighter in the mid-1980s.

The biggest export customer for the Phantom has been West Germany. Throughout the 1960s the Luftwaffe had suffered appallingly from a combination of the unforgiving characteristics of the F-104G Starfighter and its own lack of previous experience with supersonic combat aircraft, and by the later years of the decade the force's front-line numbers were falling short of the required level. The first decision taken was to replace the two *Geschwader* (wings) of reconnaissance RF-104Gs, and in November 1968 the RF-4E was selected as the replacement aircraft. Equipped with Goodyear side-looking radar and a ground-to-air data link, the 88 aircraft ordered were generally similar in specification to the Israeli RF-4Es. They were delivered from late 1971.

There was still a specific need to replace some of the F-104Gs: West Germany intended ultimately to supplant the Starfighters with a part-German aircraft, but by the early 1970s it was obvious that this aircraft – which eventually matured as the Tornado – would not be available much before 1980 at the very earliest. The F-4 proved attractive as a stopgap: a highly proven product, it was moderately priced and offered commonality with the existing reconnaissance aircraft. To reduce costs further the Luftwaffe decided to dedicate its Phantoms to the interception role, removing the rear seat in normal operations and deleting the rearmost fuel tank and items of equipment such as the bomb-release computer, the INS and the in-flight refuelling equipment. This

The 5,000th Phantom was delivered on May 27, 1978, exactly 20 years after the type's first flight. (MDC)

simplified aircraft was initially known as the F-4E(F) and later as the F-4F. The full armament of Sparrow and Sidewinder missiles was retained, along with all air-to-air fire-control features and the M-61 cannon. Slats and the strengthened wing were standard from the first day of the programme. The Luftwaffe was given the go-ahead to order 175 F-4Fs in June 1971 and a prototype aircraft was flown by the end of the year. Deliveries started in late 1972. The engines for the German F-4Fs were produced by MTU, which had built the J79 for the F-104, and MBB won a contract to turn out the outer wings, slats and tail sections for all subsequent St Louis-built F-4s, including new slats and outer wings for modified USAF and USN aircraft. Four Luftwaffe wings are now equipped with F-4Fs, and the aircraft will be kept in service until a new fighter, still in the definition stage in late 1982, is found to replace them.

The third big customer which announced for the Phantom in late 1968 was Japan. Like Germany and Israel, Japan settled on the F-4E (later standardizing on the slatted wing) but from the start of the programme decided to build both the aircraft and powerplants domestically. Only the first two F-4EJs and 14 RF-4EJs were built in the USA, first deliveries taking place in July 1972. Before that the first of eight F-4EJs assembled by Mitsubishi from McDonnell Douglas-supplied kits was flown in Japan, and they were followed by 130 all-Mitsubishi aircraft. The engines were built by IHI Heavy Industries. As in Germany, the Phantoms replaced Lockheed Starfighters.

The rest of the Phantom export customers dropped neatly into a very similar category: all three were Mediterranean Nato powers of moderate military means, and were among the few nations in the alliance to have stayed out of the F-104G programme. The first to take delivery was Spain, which received the first of 36 ex-USAF F-4Cs in 1971. Thoroughly remanufactured, the F-4Cs remain in service with two squadrons, a plan to replace them with F-4Es having failed to reach fruition. Turkey was the next customer, ordering 40 F-4Es in mid-1972. Not to be outdone, Greece also ordered the F-4E shortly afterwards. Greece received a total of 56 F-4Es and 8 RF-4Es, while Turkey's purchases eventually amounted to 72 F-4Es and eight reconnaissance aircraft. The only other operator of the Phantom has been Australia, which leased 24 F-4Es while its F-111E deliveries were delayed by programme difficulties.

Conveniently, McDonnell Douglas could celebrate the 20th anniversary of the first flight of the XF4H-1, on May 27, 1978, by delivering the 5,000th Phantom to its purchaser. By that time production was down to only twelve aircraft a month, and the closure of the production line was only a few months away. The line shut down in October 1978, having produced 5,039 aircraft. McDonnell Douglas did design a simplified air-to-air version of the Phantom, the F-4T, to meet the Carter Administration's call for an "export fighter," but the bid was not successful.

Production of the F-4 had not quite finished, however, because Mitsubishi was still working on the F-4EJs for the Japan Air Self-Defence Force. The last was not delivered until May 1981. Total Phantom deliveries stood at 5,177

**Top** Biggest single export customer for the Phantom was Germany, which started with RF-4Es and went on to buy the F-4F fighter variant. **Above** Among the last countries to take delivery of the F-4 was Turkey, whose last F-4s were among the final Phantoms off the St Louis line. (MDC)

aircraft, making the F-4 the West's most widely used fighter of the post-1945 era. The MiG-21 has probably been built in greater numbers, but the Soviet programme dwindles to less than half the size of the Phantom production effort if the vastly different size of the two aircraft is taken into account. But even in 1983, it seems, the development story may not be quite over. Pratt & Whitney is building a modified version of the F-15's engine for the Israeli Lavi fighter. Coincidentally, this powerplant, the PW1120 turbojet, fits neatly into any J79-powered aircraft while offering 3,000lb more thrust and better fuel consumption. The Phantom's durable structure (witness the continuing use of the F-4B airframes) may make a re-engine programme attractive. Applications being proposed include the F-4G Advanced Weasel, Aggressor F-4s for air combat training, and the "younger" export F-4s, particularly Israeli aircraft. The Phantom could be around – and a formidable opponent – for some time yet.

## Conclusion

Demanding requirements from the customer, chopped around and drastically altered during the development stage;

a half-way change of engine; the customer's insistence on a primary weapon system that proved to be unsuited to actual combat conditions. Any or all of these have been used as excuses for failure, but the designers at McDonnell treated them as challenges and in the process created the best Western fighter of its day. The drastically altered requirement led to an aircraft that was not only successful in the new task imposed by the customer, but also still one of the best aircraft available for the original job of ground attack. The change of engine was accomplished to such effect that the Phantom became an object lesson in powerplant integration. The PK of the Sparrow was unquestionably poor, but this was offset in service by the fact that the Phantom could carry four missiles with minimal effect on performance. The successful absorption of all the Navy's changing requirements into the Phantom was the first of the design achievements that made the fighter a classic.

The second was the defeat of the stability and control problems. They had loomed large and were indirectly a result of conflicting requirements for a heavy payload, high Mach number and short overall length. They were finally beaten after wind-tunnel testing alone – the XF4H-1 flew with all fixes in place – and by modifications which were minor compared with, say, the monstrous folding ventral fins sported by the rival Crusader III.

Third of the keys to success was the less dramatic but utterly professional job of mid-life development that produced the F-4J, F-4E and Spey-Phantom. While the Phantom had been an attractive aircraft before, the prospects of even greater payload, lower approach speeds, more internal fuel and the choice of Doppler radar or internal multi-barrel gun rendered it irresistible to any air force which could afford it. It is probably not too much of an exaggeration to say that any nation which bought any other comparable aircraft between 1965 and 1975 did so either to support a home industry, because they were embargoed by the USA or because they did not have enough cash to buy F-4s.

Excellent as the Phantom was, though, it was undoubtedly blessed with good luck as well as good designers. Its tortuous development unintentionally gave it the multi-role capability that had been carefully designed out of its contemporaries. It reached maturity just as a new regime at the Defense Department decided to resurrect the multi-purpose aircraft. The same Defense Department terminated production of all its contemporaries in favour of a new super-plane that was to replace the Phantom as well. That

In USAF service the F-4 is now being steadily replaced in the fighter/bomber role by General Dynamics' outstanding F-16 Fighting Falcon.

super-plane turned into a snakepit just as the USA became involved in the sort of large-scale conventional war which was not supposed to happen any more.

In its influence on its successors the Phantom shows the true marks of a classic. The similarity of the F-15 to the Phantom is immediately obvious in plan view, which reveals a Phantom with immensely subtle aerodynamic fixes to improve stability and control, with offset engines and sophisticated nozzles to control the base drag, with more efficient engines to extend the range. The influence may be less obvious in the case of the F-16 and F-18, but then you consider their large, fixed-sweep, moderately swept wings, high power/weight ratio and capacity for external stores and you realize that the Phantom was effectively their starting point. The philosophy behind them is recognizable as having sprung from service experience with the Phantom, its good points and its bad, and that is more than can be said of any other fighter of the day.

In brief, there is one way to sum up the achievements of the F-4: in the years from 1958 to 1972 a complete generation in Western fighter development was effectively accounted for by a single aircraft, the jagged monster from Missouri.

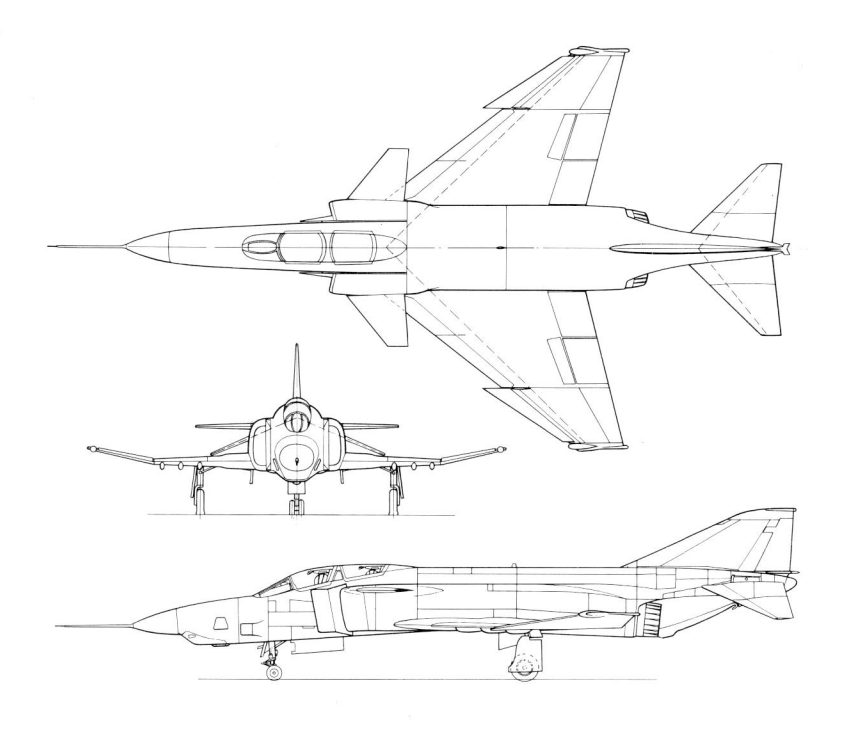

The airframe which had served as prototype for the F-4E was modified in 1972 as the F-4CCV (control-configured vehicle) fitted with an advanced automatic flight control system and movable canard foreplanes. It is now in the USAF Museum at Dayton, Ohio.

## Phantom versions

| | |
|---|---|
| **F-4A** | Designation applied retrospectively* to prototype (XF4H-1), to test and evaluation aircraft and to early production aircraft with J79-GE-2A engines (F4H-1F). 47 built 1958–60 |
| **TF-4A** | Some sources use this designation for dual-control aircraft in F-4A/F4H-1F batch |
| **F-4B** | Service-standard USN/Marines aircraft with J79-GE-8 engines. Originally designated F4H-1. 649 built 1960–66 |
| **RF-4B** | USMC reconnaissance aircraft. Originally F4H-1P. 46 built from 1965 |
| **QF-4B** | Drone conversion of F-4B, carried out 1971. Seven converted |
| **F-4C** | USAF version of F-4B with larger wheels and other changes. Originally F-110A. 583 built 1963–66 |
| **RF-4C** | First reconnaissance version, for USAF. Originally RF-110A. 505 built from 1963 to 1973 |
| **EF-4C** | Some F-4Cs converted to Wild Weasel configuration |
| **F-4D** | Improved USAF version with avionic changes. 825 built 1966–67 |
| **F-4E** | Ultimate USAF and export version with improved airframe, more power and internal cannon. 1,242 built at St Louis 1967–78 |
| **F-4EJ** | Designation of F-4E for Japan. 140 built, 138 of them in Japan, 1972–81 |
| **RF-4E** | Standard reconnaissance aircraft for F-4E export customers. 146 built 1969–78 |
| **EF-4E** | Original designation of F-4G Wild Weasel, see below |
| **F-4F** | Simplified derivative of F-4E for Luftwaffe. 175 delivered 1972–76 |
| **F-4G** | About 12 F-4Bs modified to test new communications/landing aids with operational Navy unit, around 1965–66 |
| **F-4G**** | Advanced Wild Weasel with APR-38 detection, homing and weapon-management system. 116 converted from late-production F-4Es in 1978–82 |
| **F-4J** | Second US Navy/Marines fighter version, with AWG-10 radar, improved airframe, more power and greater fuel capacity. 522 built 1966–72 |
| **F-4K** | Version for Royal Navy, similar to F-4J but with Spey turbofan engines. British designation Phantom FG.1. Half were delivered directly to the RAF. 52 built, including 2 YF-4K, 1966–69 |
| **F-4M** | Version for Royal Air Force, largely similar to F-4K. British designation Phantom FGR.2. 118 built, including two YF-4M, 1967–69 |
| **F-4N** | 228 F-4Bs remanufactured and updated for USN and Marines, starting in mid-1970s |
| **F-4S** | Some 302 F-4Js remanufactured, updated and fitted with slatted wing in the later 1970s |
| **F-4T** | Proposed simplified export version, not proceeded with |

* F4H- and F-110 designations replaced by F-4 in October 1962.   ** Designation re-used by USAF as USN F-4Gs were no longer in inventory.

## F-4 market

| User | Subtypes | Total | Dates | Notes |
|---|---|---|---|---|
| **US Navy/Marines** | 47 F-4A<br>649 F-4B<br>46 RF-4B<br>522 F-4J | 1,264 | 1958–72 | Some F-4B converted to QF-4B, 12 F-4B to F-4G. Remaining F-4B converted to F-4N. Remaining F-4J converted to F-4S. F-4 will be in first-line service until mid-1980s |
| **US Air Force** | 583 F-4C<br>505 RF-4C<br>793 F-4D<br>949 F-4E | 2,830 | 1963–76 | 36 F-4C sold to Spain, 36 F-4D to Korea, 118 F-4E to Israel. 116 F-4E converted to F-4G. RF-4C, F-4E and F-4G were major types remaining in late 1982 |
| **Britain** | 52 F-4K<br>118 F-4M | 170 | 1966–69 | In service until mid-1980s. Some 12 F-4Js acquired in 1983 |
| **Greece** | 58 F-4E<br>8 RF-4E | 66 | 1974–78 | Orders for 56 F-4E, supplemented by two replacements from USAF stocks |
| **Iran** | 32 F-4D<br>177 F-4E<br>16 RF-4E | 225 | 1968–76 | Diminishing number of aircraft, if any, now operational |
| **Israel** | 204 F-4E<br>12 RF-4E | 216 | 1969–76 | Includes 118 F-4E from USAF |
| **Japan** | 140 F-4EJ<br>14 RF-4EJ | 154 | 1972–81 | Includes 138 F-4EJ assembled or built in Japan. |
| **Korea** | 36 F-4D<br>19 F-4E | 57 | 1969–78 | F-4D from USAF |
| **Spain** | 36 F-4C | 36 | 1971–72 | Ex-USAF aircraft. To be replaced in mid-1980s |
| **Turkey** | 72 F-4E<br>8 RF-4E | 80 | 1974–78 | |
| **West Germany** | 88 RF-4E<br>175 F-4F | 263 | 1972–1977 | Replaced F-104G. F-4F to be replaced by new fighter in early 1990s. |